AF882466

Through a Cloud of Bullets

Through a Cloud of Bullets

Harold B. Willis
With Jeffrey S. Williams

Antietam Creek Press
St. Paul, Minnesota

Through a Cloud of Bullets

© 2019 by Jeffrey S. Williams

Published in the United States of America by

Antietam Creek Press
P.O. Box 11847
St. Paul, MN 55111
www.antietamcreekpress.com

Biographical Note

This Antietam Creek Press edition, first published in 2019, is an unabridged republication of Harold B. Willis's six-part series that originally appeared in the *Boston Sunday Globe* in 1918 and 1919. The Introduction, Notes and Epilogue are © 2019 by Jeffrey S. Williams.

ISBN: 978-0-9890421-3-0

This book is respectfully dedicated to all Airmen who ever served in the United States Army Air Corps, United States Army Air Force and the U.S. Air Force, and those who continue to fly, fight and win in our nation's defense.

Fight in the Air (Library of Congress)

*"Brave minds, howe'er at war,
are secret friends,
Their generous discord with
the battle ends." - Anonymous*

"Les Trois Compagnons," created in 1915 by Theophile Alexandre Steinlen (1859-1923). Part of Straight collection of French graphic art relating to World War I. (Library of Congress)

Contents

INTRODUCTION
Flying from Ourselves

In his book *Poetics*, Greek philosopher Aristotle outlined his six rules of storytelling. Great authors throughout history, including such notable names as Saint Thomas Aquinas, René Descartes, Ayn Rand and James Joyce have all been influenced by his rules. Millions of others have studied them in high school and collegiate English programs throughout the world.

His six rules governing plot, character, thought, diction, song and spectacle are just as real today as they were in Aristotle's day. When it comes to the development of character, he believed that they should be good, true to type, true to life and consistent. Whether one reads a great piece of literature or watches an exciting film, elements that Aristotle outlined so long ago are still present.

When we think of great literature or films we often look at the fictional universe. Most of these "great works" are borne out of a writer's imagination and then transferred through paper or film mediums for the enjoyment of the reader or viewer. Occasionally we will find stories that we think can only be work of somebody's imagination, only to find out that they are real, that truth is stranger than fiction. This is one of those stories.

The following story first appeared as a six-part newspaper series that appeared in the *Boston Sunday Globe* in the December 22 and 29, 1918 editions, and the January 5, 12, 19 and 26, 1919 editions and chronicled the World War I experiences of Sergeant Harold Buckley Willis.

Willis graduated from Harvard University in 1912 and worked for an architecture firm before joining the American Ambulance Service in France early in 1915, shortly after the beginning of World War I. Since America had not entered the war yet, he volunteered for the French Foreign Legion, where he was assigned to Escadrille N.124, the famed "LaFayette Escadrille" which featured American pilots who volunteered for the French Air Service.

It was through his short-lived flying career with the LaFayette Escadrille that Willis was shot down behind German lines and taken prisoner. Even though he was only a sergeant, through a bit of sleight of hand, Willis was able to pass himself off as an American officer, and was treated as such during his captivity. Then, shortly before the end of the war, he was part of a group of thirteen Americans, mostly aviators, who escaped from the German prison at Villingen, and was one of only three who successfully trudged through the rugged "Black Forest" region of Germany and made their way to the neutral country of Switzerland. All three of them obtained their freedom there.

There are two things that make this story remarkable. First, with his background in architecture, Willis was able to capture many details in his story that most writers tend to overlook. It is true that he doesn't write much about the final leg of the escape – from Häusern to the Rhine. After all, he was fatigued and sick during the last day of the trek. We cannot blame him for that. Nonetheless, the details surrounding his capture, the prisons that he was at (nine in all), and the escape, are extremely vivid in this account, as well as his treacherous crossing of the Rhine.

The second thing that makes this story remarkable is the timeframe in which it was written. While most memoirs of wartime experiences are written years or decades after the fact, Willis's narrative appeared in print for the first time only a mere nine weeks after being debriefed by American authorities in Bern, Switzerland.

In fact, as a preface to the first installment, the editor of the *Boston Sunday Globe* wrote, "When the French Liner Lorraine docked at New York last Monday there stepped briskly down the gangplank a tall, straight young officer uniformed in a way that instantly caught all eyes, even those of custom officials and port guards, who had seen so many returning heroes in varied garbs that it had all become a commonplace affair." Last Monday, of course, would have been December 16, 1918, six days before this story first appeared in print.

Exactly when or how this was written is unknown. Did Willis write this when he was crossing the Atlantic Ocean and heading for home on the *Lorraine*? Or did he grant a very long interview with a writer for the *Boston Sunday Globe*, who then wrote it into the series? While those details are unknown to us a century later, the fact remains that his story appeared six days after returning home from war and captivity, a remarkable feat when you consider the technology of the time.

For newspaper publishing during the First World War era, the story would be written by the author and given to an editor for corrections. Once the editor signed off on the finished version, it went to a typesetter who would type the words into a machine that imprinted the text onto a thin metal cast. The casts were then assembled onto a metal plate, which was run through a rotary printing press that allowed the ink to print onto each page of the newspaper. Though I am simplifying the process somewhat, it is important to understand the basics of newspaper printing a century ago to truly appreciate the speed in which Willis's story made it to the readership in the Boston area.

This is important for another reason. As an editor, I have taken the liberty of making minor changes to the manuscript due to errors that occurred because of the printing process of the original. There were numerous misspellings or typographical errors in the original newspaper series that I have corrected. Some were misspellings that Willis made during his original draft, some were most likely attributed to editors, though a bulk of them were created during the typeset process when the casts were being created. I felt that correcting the misspelled words would improve the reading experience while not taking anything away from the context or author's intent in any sentence in the book. I have also done the same with the spellings of cities and towns in order to migrate away from the limitations set forth by century-old technology, and corrected the spellings of a few names in order to maintain continuity with the historical record of events and people. Again, all of these changes are minor.

There are a few instances where I have added words, mainly in the first reference to a name. Because there are areas where parentheses are used in the original story, I have elected to denote my editorial changes inside of brackets instead.

In order to maintain the flow of the story as written and to keep with the author's intent, the punctuation and subheads remain as they were written.

Since this was written at a time when the readership understood the places, people and events of the First World War, there are areas where further context is needed for the modern audience. I have added footnotes with brief explanations to enhance the narrative without making changes to Willis's original text.

Even though I have chosen not to add a bibliography for the notes, there were a few sources that were invaluable as references. The books *Prisoner of the U-90* by Edouard Izac; *Lafayette Flying Corps* Vol. I and II by J. Norman Hall; and *Escape from Villingen, 1918,* by Dwight R. Messimer, were instrumental in this effort. Additionally, "An Airman's Escape," a two-part series in the April and May 1919 editions of *The Atlantic Monthly* by George Puryear, and Steven A. Ruffin's article "Major Harry Brown and His 'Lost Flight' of the 96th Aero Squadron," published in the Fall 2004 issue of *Over the Front* by the League of World War I Aviation Historians, were also most helpful. The World War I records of the International Red Cross are digitized and online, which enabled me to verify information on some of the prisoners held at the camps, including their dates of repatriation. Those can be accessed at https://grandeguerre.icrc.org/ It is doubtful that this story could be as complete without them. Not only did these works greatly assist in creating the notes, but they also provided independent verification regarding the historical accuracy of Willis's experiences.

Since Willis left the camp at Villingen, he had no way of knowing what happened to the rest of his comrades who attempted the escape by the time his story went to press, and consequently left the story of the others unconcluded. I have added an Epilogue as a final chapter in the narrative to tie up these loose ends. Most of the information was provided in Dwight R. Messimer's *Escape from Villingen, 1918,* as well as from obituaries and other records.

Also, keeping in line with the headline style of the day, the series had no formal title. Therefore, I have given it the title *Through a Cloud of Bullets*, which is taken from one of the subheads in Chapter Two, and originally printed in the December 29, 1918 edition.

When I first read this story, it brought back memories of my own experiences while serving in the Air Force in Kirkuk, Iraq, during Operation Iraqi Freedom in 2003-2004, and as a wildland firefighter trudging through the rough terrain in Eastern Oregon in 1996. I was just a little bit older than Willis during each of these experiences.

Despite the passage of time, the improvements in technology and the differences in geography, the experience of Willis, minus his confinement, wasn't too unlike mine, that of my father who was a Marine in Vietnam, or my grandfather, a Marine in the World War II Pacific theater.

In the words of Erich Maria Remarque, the German World War I veteran who penned *All Quiet on the Western Front*, "We are not youth any longer. We don't want to take the world by storm. We are fleeing. We fly from ourselves. From our life. We were eighteen and had begun to love life and the world; and we had to shoot it to pieces. The first bomb, the first explosion, burst in our hearts. We are cut off from activity, from striving, from progress. We believe in such things no longer, we believe in war."

Remarque was only eighteen when he went to war. Willis was a few years older, twenty-five, when he first arrived in France, and had reached the mature age of twenty-eight when he wrote this piece.

Finally, my deepest thanks go out to my friend and fellow Air Force veteran Christian Michael of Scroll Media/Icon Covers for his excellent work in designing the book cover; and to my mother, Barbara, for her support during the editing of this project.

Chapter One

Our Fourth of July Dinner

The magnificent gain of territory by the American troops which cut off the St. Mihiel salient, that thorn in the side of the French Army, recalls to me the bitter struggle made by the French to cut off that promontory of German territory in 1915.[1] The salient was deep and narrow; the obvious stunt was to attack at the base and cut it off – like the claws of an ice tongs.

Our little American Ambulance Section No. 2 alternately enthused and despaired as we did our little bit to help France push in one part of the tongs in the Bois-le-Prêtre.[2]

That was long before the time of week-long artillery preparation and the massed attacks on 30 and 40-kilometer fronts.

The tactics of General LeBoque, who commanded our division, the 73rd, rather resembled the surprise attacks developed during the last stage of the war, excepting that he had no tanks, only a few batteries of 75s and some old-fashioned heavy artillery which had been discounted from the forts of Toul.[3]

A round dozen of us arrived in that quaint town of Pont-à-Mousson, which is well known to all American troops who have held positions on the right of our front. We had rolled all the way, the dusty way from Paris, in Fords more or less made over into ambulances, with canvas prairie schooner tops. Everywhere we were greeted as the first Americans to be seen on the French front, and those who know the people of Lorraine need not be told that our welcome was a warm one.

Not longer than three days after our arrival, before the novelty of the whine and crash of shells had passed, the flower of the 20th French Army Corps commenced its magnificent but futile attacks. Morning after morning we would be awakened before the gray of dawn by brief half-hour preparations. Then the pop and rattle of musketry which precluded a day of breathless activity transporting those pitiful, bleeding wrecks of men to the

1 The September 12-15, 1918 Battle of St. Mihiel was still fresh in the minds of newspaper readers when this series premiered in the *Boston Sunday Globe* on December 22, 1918.

2 Richard Norton organized the American Volunteer Motor Ambulance Corps in the fall of 1914 shortly after the start of the war. As he was a graduate of Harvard, class of 1892, Norton reached back to his alma mater for early recruits. Harold Willis graduated from Harvard in 1912 and was assigned to Section Sanitaire Americaine No. 2, stationed in Lorraine after he was recruited.

3 French Army's 73rd Infantry Division. Toul is a municipality in northeastern France that featured stone ramparts designed by King Louis XIV's military engineer, Sébastien Le Prestre de Vauban. They were designed in 1698 and built within the next two years. From 1910-1914, the French built twelve additional smaller forts, plus 27 smaller works and three redoubts. They were armed with 402 guns, including four 155-mm artillery pieces inside of two Mougin turrets, and twenty 75-mm artillery pieces.

operating section in the rear.

Then we came to learn how Frenchmen never groan, though enduring unspeakable agony, how they would apologize for the moans of a comrade who had become delirious with pain, how they would look with grateful recognition in their eyes when we could in any way ease their sufferings.

Gradually we would find out the results of the day's work – a few hundred yards of trenches won, in the tangled labyrinth of the Bois-le-Prêtre an annoying blockhouse destroyed.[4]

"Ah, my old one. A few more days like this and we will have them out of the woods, over the top of the hill. Then you will see them run like rabbits and we will be before Metz."

The heroic work of these young soldiers did push them out of the woods and over the hill. But then it was found that once over the hill one was in a very unfavorable position. First, the Germans were blown off the crest of the heights; then the French, then the Germans; then the French again.

Meanwhile the terrible attack through the mud at Les Éparges had won a village and a hillside. But St. Mihiel seemed to be as solidly affixed to the Hunnish territory as before.[5]

Our Fourth of July Dinner

The shock troops were withdrawn from our little divisional front. Either through spy information or aviation the Hun was able to take advantage of that critical moment when our active Army troops were being withdrawn.[6]

On the Fourth of July we were called for as we were commencing to taste the soup of our elaborately prepared dinner. A German artillery preparation was starting. A call came from the advance dressing stations for every available man and car. Our Fourth of July dinner was finished during

4 Bois-le-Prêtre, translated as "Priest's Wood," was a WWI Theater of Operation from September 1914-July 1915 involving the French 73rd and 128th Infantry Divisions against the German 121st Infantry Division in the Lorraine forest near Pont-à-Mousson. It was one of the first missions for the American Ambulance's Field Service.

5 The town of St. Mihiel formed a salient inside the French lines, blocking communication between Nancy and Verdun. After failing at Bois-le-Prêtre, German troops took St. Mihiel in the Battle of Flirey. They would hold it for the next three years until U.S. Army General John J. Pershing ordered an attack on September 12, 1918.

6 "Hun" is a slang term used in reference to the Huns who conquered Eastern Europe around 370 A.D. The term was taken from a speech by Kaiser Wilhelm II on July 27, 1900 when he compared the modern Germans to Huns for making names for themselves in wartime. The allies used this as a pejorative term for all German soldiers during WWI. Willis uses this term continually through the remaining narrative.

the day whenever came the pauses necessary to refill our gas tanks giving us opportunity for a hasty bite and a sip of champagne.

Toward evening of that attack the Germans gave two of us who were fortunate enough to be at the head of the line an opportunity for our first thorough shooting up. The call came for a couple of cars to go up through the French part of the Bois to the reserve lines, where some officers with head wounds were waiting to be transported to the operating table.

LIEUT WILLIS RIDING IN BACK OF HIS CAR AT VERDUN

This photo appeared in the December 22, 1918 edition of the Boston Sunday Globe. Willis is on the right.

The jolting of regular means of bringing them to the dressing stations on stretchers would have been fatal to them. So we jolted up the rough wood in comparative quiet, with only the "ts-ing, tsing" of spent bullets about us to give us occasional thrills.

No sooner had we reached the dugout when the row commenced. The Huns laid down a barrage just about where we had stopped in order to prevent reserves being brought up.

I remember the dugout where we lay had only a corrugated sheet iron roofing with a little dirt laid on it, and the 30 poilus who shared it with us had only a too vivid realization of its frailty.[7] In the heat of the worst of it, with shells crashing into trees about us, with splinters in all directions, one of the French stretcher-bearers walked to the entrance of the dugout to tell us that the wounded were ready.

His fearlessness rather shamed us, but it was the hardest thing in the world to go out and stand in the awfulness and unprotection of outdoors. First came the customary wrestling match with a new-fangled type of stretchers. Then we went out for our wounded, lifted them over the edge of the trench and slid them into the cars, wondering how much longer we would be permitted to stand in that hail and crash of shell splinters without being laid low.

Luck was with us. For once the engine started on its first cranks. One of my cylinders started missing as I climbed into the seat. I found later

7 Poilu is an informal term for a French infantryman in World War I.

that the ignition had been cut, but I probably made the quickest turn around on record, two twists and a couple of kicks. Then we couldn't go down that road fast on account of the deep ruts and the delicate condition of our patients.

Every once in a while pieces of metal clanged against the metal of our machine or ripped through the wood and canvas behind us. Why our inert burdens were not again wounded is beyond me, but we finally got clear of the barrage in the woods, with 15 or 16 holes gashed in each of our cars.[8]

Photograph shows an American ambulance driving through the Woevre Forest, France during World War I. Most ambulances at this time were built by the Ford Motor Company. (Library of Congress)

In the Depths of Despair

Only once or twice did I have closer calls in the air.

Among the untried Americans despair filled every heart. Our first lines were gone. The Huns were advancing through the wood and down the hill. That precious ground to gain which we had seen so many thousands march up the hill never to return was lost. The splendid lives had been paid in vain.

---

8 Willis's actions during this attack earned him a citation from the French military.

Although the Germans had gained only a few hundred yards, to us the little sector had come to mean the whole war and for weeks we thought that the whole of France was lost. We had so little comprehension of the real scale of the war. The depression in some of our letters even brought criticism from the censor.

Meanwhile the Hun had neglected nothing – not even his warfare on women and children. When we arrived at Pont-à-Mousson, the town was intact. Children were going to school. All the little shops were open. People attended mass as before. Soon after our arrival the cool, systematic bombardment of the city commenced, special attention being paid to the great central marketplace, where the life of the community centered, and to the hospitals.

We had our quarters at the extreme end of the town and our work consisted of bringing wounded from the dressing stations back of the trenches to the hospitals, located in different sections of the town, whom we later evacuated them to the rail head. Bombardment could not interrupt the service and so we made a serious study of the system of bombardments. Certain streets received more shells than others, we noticed. Going through a town, if we heard a shell drop in one street we would always turn up it, for the Hun artillery swept this target across in a methodical manner. I remember some of us kept maps and marked the hits to discover if a definite pattern was used by the Hun gunners. But there were too many hits to record.

After the bombardment was over came the most painful of our services, to go about picking up the women, the men and those little children that we could find lying in little crumpled heaps on the blood-stained cobblestones.

Is it strange that all of those original 12 in our section of the American Ambulance were fighting in the French Army a full year before America showed signs of taking positive action in the war?[9]

During the hours when we were not on service we would take long walking trips up to the trenches to visit our French chums, to the 75M Battery positions, where we would marvel at the difference between their breech mechanisms and those of Battery A, Boston.[10]

More instructive still were our excursions across the river to the eastward, to the battle-scarred convent in Mousson, which still served as a military hospital, with the kindly, sweet-voiced nuns in charge, despite its hundred bombardments; up to the observatory of the old castle on the heights

9 The United States didn't enter the war until April 6, 1917.

10 Though he doesn't provide clarification, Willis is most likely referring to Battery A of the 101st Field Artillery Regiment.

of Mousson, where, through the powerful telescopes of the artillerymen, we could watch Boches[11] raking hay, milking cows and riding bicycles far back of the lines.

German "Fun" a Reward for Victory

The greatest trip of all was to Nomeny, 10 miles to the eastward, where the Hun had left tracks of his cloven hoof to be seen by all the world.

Later, German officers explained to me – in captivity – why the discipline had to be relaxed after battle and the soldier given his fun in reward for the hard-earned victory – "fun" in which enemy women and children and private property paid the price.

In General Castelau's retreat to a successful position about Nancy, Nomeny had been given up.[12] Its crumbling walls today show no mark of shell or bullets, except where a swinish and drunken soldiery shot down civilians as they crept out of entrances to their cellars when homes were fired

Photograph shows an American ambulance on the road to Verdun, France during World War I. Note the exploding shell in the distance. (Library of Congress)

11 Boche is a French contemptuous slang term for a German soldier, meaning "hard head" or "head of cabbage." The term was used during both world wars. Willis interchanges this with the term "Hun" throughout the narrative.

12 General Noel Edouard Marie Joseph, Vicomte de Curieres de Castelnau was commander of the French 2nd Army.

above their heads. The men who showed us about were the same men who had picked up and buried the bodies of the victims.

With tears in their eyes they would point where this little family group had been found, where those two children were buried in the middle of the street – where the mother of the French major whom we had known had been shot down on her own doorstep.

Everywhere the broken glass of beer and wine bottle told its own tale. "Ah," they would say, "if ever one of those dirty ones comes between these two hands of mine--" and a gesture would finish the sentence.

All this was teaching us to yearn to finger the throttle of a Nieuport flying fighter in place of the trusty ambulance.[13]

It was in the first months of 1916 that the efforts of Norman Prince to form an escadrille of American airmen had been crowned with success.[14]

When the French Government sent us to Verdun, just before the beginning of the Great German drive, we found ourselves in the same sector with the first little American Lafayette squadron – those heroic figures who made such a splendid name of themselves in the early history of the war - William Thaw, Elliott Cobb, [Victor Emmanuel] Chapman and [James] McConnell, the first who had left our little ambulance section to go into aviation.

[Charles] Chouteau Johnson, Clyde Balsey and the founder of the escadrille, Norman Prince, those men were over our heads every day. Frequently they landed to take lunch with us. By chance, when Thaw was shot down, with his left arm shattered, and Norman Prince was forced to make an emergency landing, some of us were called to pick them up.

The temptation to take a fighting job in the Great War was too strong to be resisted. One after another we left to enroll in the Foreign Legion, from which we were immediately transferred to French flying stations.[15]

Walter Lovell of Boston (who later became a captain in the American aviation forces) and I left together, the first of June, 1916. [Willis] Haviland, who won a great reputation in the United States Navy flying force; John Huffer, now commander of a pursuit group in an American squadron; and Kenneth Marr (now Major Marr), one of the first in the list of most efficient pursuit squadron commanders, left about the same time.

---

13 Nieuport was a French aircraft manufacturer from 1908-1937.

14 Escadrille is French for "squadron."

15 French Foreign Legion was established in 1831 and uses foreign recruits for its soldiers.

So many stories about aviation schools and training and learning how to shoot in the air and doing acrobatic tricks, that there is no need to bore you with tales of our experiences. We had the great good fortune to be taught to fly in monoplanes at the old school at Versailles. It was slower work, but interesting. Of course, with the unreliable engines the forced landings came often, and we would always take care to make them near agreeable chateaus.

One of my pleasantest souvenirs is a splendid welcome the French gave me every time I was forced to come down with more or less trouble. Once I spent a week in the cottage of a peasant whose hospitality was as warm and as much appreciated as that of the most elaborate estate.

We got to the front in time to see the last one of these American airmen we had known at Verdun, Chapman. [Kiffin] Rockwell and Prince had already paid the big price.[16] [Elliott] Cowdin and Balsey had been forced by illness and wounds to relinquish their grasp of the "joystick" as the control is called.

We had not flown more than a month before our old comrade of the early ambulance days, McConnell, went down during the first great German retreat on the Somme.[17] Then we came again to see the fiendishness of Hun treatment of civilians, his wholesale pillage and thieving which one could understand; his wanton destruction of most insignificant personal property which he did not care for was inexplicable. Acres of poor little apple trees slashed down; wells poisoned; chateaus burned at the last minute before leaving.

During the Spring and Summer months we were moved from one part of the front to another, first to the south for the big attack on the Aisne, where we had our first glimpse of the bleached bones of Reims, the Chemin des Dames and the Champagne front. As we were an independent group of fighting planes attached to no army, when the big English push was started at Ypres we took a flying field on the beach at Dunk to fly with the 1st French Army Corps, which acted as the left wing of the English drive.[18]

Then we made the acquaintance of Freiherr von Richthofen and his grotesquely painted red and black squadron, and the more pleasant acquaintance of our brother aviators in the English Army and Navy. Immediately

16 Kiffin Rockwell died during a dogfight on September 22, 1916. Norman Prince died in a crash landing October 15, 1916.

17 James Rogers McConnell lost his life in aerial combat against the Germans over the Somme on March 19, 1917.

18 Third Battle of Ypres lasted from July 31 to November 10, 1917. "Dunk" is probably Dunkirk.

they became our playmates during those pleasant hours, which it is the aviator's good fortune to enjoy between flying time.[19]

Unexpectedly, one day, the order came to pack our duds immediately and leave ground in the direction of Verdun. All of us had a premonition that this move spelled some hard fights for us and certain losses. For Verdun had been the grave of countless airmen as well as of poilu.[20]

Our fears were not unfounded. When we landed on a little field in the Argonne we found that our work consisted of protecting daylight bombardments down the Meuse, the most difficult and dangerous work for the pursuit squadron which had already cost the escadrille Norman Prince, and which cost it the more or less useful services of the writer the very first time we crossed our lines in a big bombardment.

Our first two or three flights to look over the country showed us the Huns were there in force. The afternoon before our attack I remember [Raoul] Lufbery came down with his plane so riddled with bullets that it could not be flown again, and two or three others of us had our wings perforated with bullets, uncomfortably close to our heads.

However, one knows in the air that one's day must come sometime. One takes it for granted and doesn't think about it.

So the next morning – the 18th of August, 1917 – we tumbled out of bed with the gray dawn to accompany a big formation in the direction of the Crown Prince's headquarters.[21] We talked and thought about nothing except getting back in time for a tramp in the Argonne, planned for that afternoon, to gather mushrooms for our mess.

I little thought that my next meal would consist of sour, soggy black German bread and stinking German cheese, with a group of stiff, monocle Hun officers with shaven pates and dangerous questions, for my tablemates.

---

19 Manfred Albrecht Freiherr von Richthofen, better known as the "Red Baron," was the most notorious German fighter pilot of WWI, having shot down 80 allied aircraft. He was shot down and killed on April 21, 1918.

20 The Battle of Verdun, fought from February 21 to December 18, 1916, was the longest battle of the war to that time, and had nearly 800,000 total casualties including an estimated 300,000 killed and 216,000 wounded.

21 Friedrich Wilhelm Victor August Ernst, the eldest son of Kaiser Wilhelm II, commanded the Army Group Crown Prince from 1915-1918.

One-half right front view of four Nieuport 17s of the Lafayette Escadrille; probably in France, circa 1917. Tail number N 1993 flown by Sergeant Walter Lovell of Boston, Massachusetts, can be seen on the closest Nieuport. (Robert Soubiran/National Air and Space Museum Archives)

Chapter Two

Banking, Twisting, Spinning

Before describing the last air battle that I took part in, which resulted in my forced descent in German territory, it might be well to say a word about the armament of our machines and the tactics that we employed at that time. As all diligent readers of war stories know, the one-seated fighting airplane, like our Nieuports or Spads, have for weapons of offense two fixed machine guns which are solidly bolted to the frame of the machine and can only be shot forward in the direction in which the machine is going.[22]

The guns usually shoot through the propeller and there is a synchronizing device which connects the engine with the gun, in order to prevent the bullets hitting the blade of the propeller. One aims by aiming the entire machine, which is easier and steadier than you might think.

Of course this arrangement means that our method of defense is a sharp, decisive attack. You have no gun shooting backwards and no armor on the plane. When, on account of a gun jam, motor trouble or overpowering odds, you have to take to your heels, you have absolutely no means of defending your tail.

The morning of my last flight over the lines – Aug. 18, 1917 – we had the mission of defending a big group of bombarding planes which were going north from Verdun in the direction of Mézières and Charleville, the headquarters of the Crown Prince.[23] The big bombarding planes climbed slower than we did, so that we allowed them to leave the ground three-quarters of an hour before we got away with our more powerful little Spads.

Ted Parsons of Springfield, [Massachusetts], who is now counted among the most successful of the American Aces, and I left the ground together, and gained our altitude in great spirals side by side. We met the bombardment formation at the appointed minute over our rendezvous above the corner of the Argonne Forest, and easily took our stations.

A patrol of fighting planes defended each side of the bombardment group and a stronger patrol took position above and behind. I remember as we passed over the great aviation center and base hospital of Vadelaincourt we saw a certain amount of dust and smoke blowing about. Months afterward I learned from a later captive that the Huns had shelled the hospital

---

22 Societe Pour L'Aviation et ses Derives, known as SPAD, was a French aircraft manufacturer from 1911-1921.

23 This flight was in support of the Battle of Langemarck, fought between August 16 and 18, 1917, during the Third Battle of Ypres campaign.

with a big marine piece in the Woëvre and had killed several doctors, nurses and wounded men.[24]

Great plumes of smoke and the puffs of exploding shells that we could see, but not hear, told us that a lively attack was in progress under cover of a gas cloud in the Mort Homme region. That was the last thing I had leisure to notice on the ground, until I was forced to land far back of the Hun balloons, which were nervously drawing in their horns at our approach.

The anti-aircraft, or "archies," as the English call it, gave us a fairly warm welcome, but as there were nearly 40 of us they could only shoot in the mass and did no harm. Not so the German aviators. Hardly had we crossed the line when the patrol on the opposite side of the bombers from us began to get action.

A Hun one-seater was indiscreet enough to attack a bomber in front of Captain Walter Lovell of Newton, [Massachusetts], and was almost immediately shot down in flames. A couple of other Huns succeeded in setting fire to one of our big bombing planes, which dove sharply and fell, turning slowly, leaving behind it a streak of black smoke.

Meanwhile, on our side, we were keeping our eyes open for Huns and on each other. The two flanking patrols had been ordered by an officer who was replacing our regular commander not to leave their position in relation to the bombers. On account of our superior speed we had to zig-zag quite violently to stay back, and one had to keep one's eyes open, not to run into other machines. Our inability to maneuver as a patrol, to defend ourselves with our customary attacking tactics, was the first cause of my disaster, and nearly got Lovell as well.

By the time we were at some 25 or 30 kilometers in German territory the Huns received reinforcements to the extent of about 10 new single-seated Albatros planes.[25] These immediately fell on our backs. One very audacious Hun dropped right in the middle of my patrol and started to drop Bob Soubrian, who was in front of me. He was easy meat. I let him have about 150 shots before he turned over and went down in a spin. I should like to find out whether I really got him, or if he was merely fed up.

24 The Bombing of Vadelaincourt occurred on August 20, 1917, two days after Willis was shot down. However, the supply station had been shelled by artillery in the preceding days. Since the actual bombing of the hospital was by German aircraft two days after Willis was shot down, he most likely witnessed a preparatory artillery shelling, not the actual bombing.

25 Designed by Robert Thelen, the Albatros D.III fighter was built by the Albatros-Flugzeugwerke company and began its service in August 1916. There were 1,866 built and it was flown by several of the top German aces including Manfred von Richthofen and Kurt Wolff.

Banking, Twisting, Spinning

Hardly had he disappeared from sight when my troubles commenced. That instant I found myself the last machine in my patrol. No one behind to protect me, and a Hun setting down on my tail.

With the first pop-pop-pop of his bullets around my ears I decided that rather than stay with the patrol and be drilled through and through before half a minute was up I would turn and mix it up alone. We had superior machine and engines, which enabled us to outfly and out-maneuver the Hun. Had he not gotten a shot in my carburetor somewhere with his first burst I would have had a very decent chance to get him, or at least to save myself.

A vertical turn and I was under his tail. I pulled open the throttle to climb out above him and get into position and – "paff," the engine checked, and almost stopped. My machine was in a steep bank and instantly fell in a spin. When I came out I had lost a 100 meters or so of precious height.

The enemy had lost me when I went under him but soon he spotted me and came down with a swoop, his two guns going full blast. From then on it was a losing fight, I banking, twisting, spinning to keep him from getting behind my tail and finishing me off.

Every few weeks or so one of your comrades goes out on patrol and never returns. One by one the old boys just disappear and new faces replace them. One knows that some morning he is going up into the blue and will not be back to take his place at the dinner table. You take all this for granted and don't think about it, except perhaps the mild curiosity as to how it will happen and what it will feel like.

When I found myself in this hopeless position my first thought was that my day had come. It was about due. For I already had a long air life – nearly a year and a half. To exasperate me occasionally my good old Hispano Suiza would have a new lease of life and allow me to climb 40 or 50 feet, only to check and sputter again.[26]

For a few instants I could follow the big group of my friends continuing on their way – a couple of Spads straggled right over my head, but failed to see me, and then they disappeared, leaving the sky freckled with smoke from anti-aircraft shells to mark where they had passed. I saw two other Huns going by my special enemy, but they merely circled above us and let him play with me.

26 Hispano-Suiza was a Spanish automotive manufacturer that also made aircraft engines. The Hispano-Suiza 8 was the most commonly used allied aircraft engine with over 49,800 built from 1914-1918.

Through a Cloud of Bullets

Finally he gave up trying to get behind my tail and contented himself by sweeping me with his fire from the side. He did good execution. As I would pass through his line of fire, although I could not hear his gun, I would hear the pop-pop-pop of his explosive bullets as they would come nearer and nearer to my head. The last one or two would clank into the engine or snip off a piece of the gun or plane in front of me. Two of my cables were cut and fluttered back in the wind.

A ball half cut off my windshield, which was luckily not of glass. Another shattered my glasses and nicked a piece of my scalp. That was like a blow of a hammer, and stunned me for several seconds. It must be the last blow.

Sergeant Harold B. Willis, Lafayette Escadrille, in 1917. (John Guy Gilpatric/National Air and Space Museum Archives)

Then I was conscious of light and noise about me again and I was still in the air in the same old machine. I couldn't believe that I was alive. Not far below me were the old rose and mauve-colored roofs of a French village, on the banks of the meandering course of the Meuse. It was German territory.

Those houses were full of Huns. If I landed in those fields I would be in the power of the Huns for the rest of the war.

Ah, that was a bitter moment. For an instant the temptation came to put my nose down vertically, close my eyes and await nothingness. Then common sense or one's animal instinct to cling to life told me better, and I looked around for a landing place, not greatly worried whether I crashed or not.

As I approached within a couple of hundred feet of the ground my enemy ceased his fire, whether from humanity or lack of ammunition I do not know. I will give him credit; he fought like a gentlemen.

There was a wood directly below me that I hopped over with a twitch of the stick. Then I glided down toward a clear space, pulled up and touched, wheel and tail together – the best landing I ever made, and on the bare summit of the crest above Dun-sur-Meuse.[27]

In a dazed way I looked up and saw the Hun wave his hand as he went over my head. He made a turn and landed below me, accompanied by his two friends. They ran up, saluted, addressed me in Sam Bernard French, asking who I was, what my machine was and so forth.[28]

I hoped that he was at least a German ace. But, no, I was the second machine he had shot down.

Buries His Orders

Soldiers, afoot and mounted, came hurrying up from all sides, but the three aviators stayed around me and prevented rough handling and robbing that I certainly would have undergone otherwise, judging from the experiences of other aviators shot down whom I have known.

Would that all the Huns in whose power I was during the next 14 months had the same respect for the rules of war, fair play and plain humanity that those young German pursuit fliers showed. Note I carefully exclude bombers.

Soon a great circle of about 100 soldiers formed around me. But the presence of the officers kept them in order. Meanwhile one of the Hun pilots was taking out my instruments and maps, which, of course, contained no information about our side of the lines.

We looked over my old boat. She had caught a lot of stuff. Two of my wing cables were gone, both tires deflated by bullets, eight or nine bullets in the engine, and I don't know how many in the fuselage about me and in the wings. When I felt down in the mouth during the days of misery to come I tried to think of the way that machine was shot up and how lucky I was to be alive at all.

It seemed to be the thing to look over strange machines, so I asked to be allowed to look over the Hun machines. My enemy agreed and took me over. It was D-3 Albatros, with two parabellum guns, a perfectly oval body made out of veneer. Compared with my machine it was crudely finished.[29]

On the way back I wadded up a couple of transport orders that I had in pocket, which might indicate to the Huns what field I had come from

27 Dun-sur-Meuse is a municipality in the Meuse department in northeastern France.

28 Sam Bernard was an English-born American vaudeville comedian and early silent film star at his peak of career during WWI.

29 Most Albatros d.iii aircraft were equipped with IMG 08 machine guns manufactured by DWM. Karl Heinemann redesigned the IMG 08 for better use in aircraft and came up with the Parabellum MG 14.

and pretending to drop a glove I pushed the paper into a mole hole. I think I got away with it.

When we returned to the crowd the commander of the aviators had arrived, and he and I sauntered down toward his quarters, which were nearby. There wasn't any handcuffing. It was really quite informal. We chatted about things not connected to the war and the only thing that indicted I was a captive were two mounted Uhlans who followed at five paces, with their lances lowered and pointing at the small of my back.[30]

Of course, it wasn't a moment to start anything.

They then led me down to Liny, and as I approached the miserable stone houses of the village my nose was greeted with the sour odor of black bread, sauerkraut and cheese that characterizes the German soldier and all he eats.

There was some talk of taking me before the German division commander, but their plans were changed, and I was led into the room of the squadron commander, where I was given a comfortable chair and a very decent breakfast of black bread and butter, sausage and fine beer. It was a heartily eaten breakfast, too, because I had nothing to eat since starting out at daybreak, and it was now about 9 o'clock.

Only once again during my captivity did I have such good beer and such an excellent meal. There was a spirit of comradeship about their whole behavior, and it was simple enough to see why. They wanted to get me to talk.

Other officers joined in the conversation, all of them arrayed in elegant uniforms, beautifully tailored and monocle, and all with shaven heads. They spoke French well and their conversation would run something like this:

"We, you know, leave the ground in such and such a formation. We start on our missions so and so. Don't you think that a good method?" And such questions designed to draw from me comments that would disclose the French air tricks. They would talk in German among themselves and then start the questions to me again.

Occasionally pointed questions would come and I would have to refuse point blank to answer. But this they did not at all resent. They had other means of getting answers which would clear up the whole matter later on, they thought.

Finally, after a whole forenoon of that, luncheon was served and I received the same food that they did, although I did not eat with them, of

30 Uhlans were Polish light cavalry regiments armed with lances, sabres and pistols. In 1914, the Imperial German Army included 26 Uhlan regiments. During WWI, British and French soldiers began to refer to all German cavalry as Uhlans.

course.

We had scarcely finished when an officer of the German Intelligence Corps came in and said that I should go with him. He was Captain Hardy, spoke perfect English and told me that he had lived for 15 years in England. We started out in a run-about of his, about three-horse power, I should judge. It had just room enough for us to squat in and just power enough to drag it over the ground; and it rolled us toward the Fortress of Montmedy.[31]

Four members of the Lafayette Escadrille are surrounded by a group of men, probably in France, prior to March 1917. From left to right are: Walter Lovell, Edmont Genet, Raoul Lufbery, and James McConnell. (Photo by Robert Soubiran/National Air and Space Museum Archives)

I kept my eyes open, hoping to see things that might be of value if I should be able to get news back to our lines. I could note ammunition dumps and airplane devices before we finally came into the town just at the foot of the fortress.

The first thing that struck me everywhere behind the German lines was the fact that every Hun, officer and man alike, had a black cigar stuck in his face, not very good cigars, to be sure; but a cigar.

Captain Hardy hopped out of the ramshackle old car and ran up to the intelligence office, leaving me guarded only by his chauffeur. I spied some old French women standing about, and they were sly ones. They would walk up in the rear of the car, where the chauffeur could not see them, and, without looking at me, would whisper questions at me.

"What was the situation? Was there hope? Who was I?" So they whispered, and I was glad to tell them that I was an American and to tell them that more of us were coming, millions more. Later they did much to help me in that fortress.

I was led into the wretched prison of the fortress, where the first hardships were encountered.

31 The Citadel of Montmedy was built by King Charles V in 1545 to replace an earlier castle that was built on the site in 1221. It was also the destination for King Louis XVI and his family when they fled Paris during the French Revolution in 1791, though they never arrived. It is one of the largest fortresses in northeastern France.

Willis Starting on His First Flight in a Bleriot, the Model That First Flew the English Channel

This photo of Willis making his first solo flight appeared in the January 5, 1919 edition of the Boston Sunday Globe.

Chapter Three

The Trick Chamber

It is hard to describe the cold despair that one feels on hearing the clank of the bolts and chains when the door of the first prison is slammed behind one. I found myself in a curious old cell with the heavily barred windows, the masonry walls, the unused fireplace and the crude furniture of the conventional stage setting. The old Louis XIV prison was located within the walls of the citadel which dominates all the country about Montmedy.

The thought came to me that I would probably never be able to fight again at the front; that from now on I would be a "Dud," the source of endless worry for my family and friends. I confess that then I lost my nerve for a moment.

Suddenly the bolts of my door were clanked open again and a rather uncouth-looking officer burst in and commenced to sympathize with me most effusively – too effusively for my state of mind at that moment.

Instinctively I disliked him on account of his dirty, unkempt condition and his little pig eyes, too close together. He did not look square and we found out later that he was not. Seeing that I was not in the mood for reminiscences he left me. Dinner was brought, consisting of a hard white brick of cheese which I promptly put out on the window sill to eat with great gusto some days later, and a hunk of black army bread.

In Power of Boches

This German "pain" is probably the most curious food that human beings ever tried to bolt down. The type made for the army had the consistency and weight of heavy wet clay – it stuck to your teeth and roof of your mouth. The flavor is of an indescribably sourness and one comes across straw bran and sawdust. I could not eat it and went to sleep on my straw mattress supperless.

Aside from the emergency dressing of my head wound at the German aviation field I received no medical attention. Happily it proved to be a very superficial graze, though it frightened me badly enough when I got it.

Before entering the prison, while my guards' back was turned, one of the old women of Montmedy had told me that soap was not to be procured in the prison for love or corruption. I insisted to Captain Hardy of the

German Intelligence that he procure me a piece before he had me thrown into the prison. He finally agreed and bought me a piece of pink and yellow composition vaguely resembling soap which was to serve me for my washing and laundry for the next two months.

The awakening from a doze was the second most terrible moment of my captivity. One heard the birds, sensed the sunlight streaming in one's room, but did not realize at first where one was. It seemed too quiet for the Escadrille, there was no roar of motors being tried out, or the swish and whistle of the return of the early patrol as they skimmed over the roof of our camp to land for breakfast. Then rubbing one's eyes, one realized that it was a cell-prison-captivity – in the power of the Boches. I experienced a terrible sinking sensation.

Other Prisoners Enter

At about 8 in the morning two blackened and bandaged French officers were let in to see me. They had had a most miraculous escape, only equaled by some of those of Captain J. Norman Hall's. Hall, as Boston folks know, had as striking a series of war experiences as any aviator. He left Boston before the war started and was bicycling through England in July, 1914. He jumped in at once, serving through the first hopeless campaign as a "Tommy" in Kitchener's mob.[32] He came back to America in 1915, but couldn't stay on this side.[33]

He joined the Escadrille, had a terrific air battle with eight Huns; escaped, though injured; went back to flying; was shot down in Germany, badly injured, and later escaped.

The two battered French officers told me their story. At about the same time that I was having my bad quarter of an hour, their bombing machine (part of our formation) was set on fire by a Hun. Lieutenant Aignan, the pilot, enveloped in flames and smoke and blinded, cut his engines and headed down. The wretched observer, hoping to save himself by a last desperate effort, unbuckled his belt and crawled along the fuselage of the whirling, spinning machine and, in spite of the heat and smoke, attempted

32 British Secretary of State for War Herbert Kitchener created a 500,000 soldier-strong all volunteer army in August 1914. According to J. Norman Hall, "The seasoned regulars of the first British expeditionary force said it patronizingly, the great British public hopefully, the world at large doubtfully. 'Kitchener's Mob,' when there was but a scant sixty thousand under arms with millions yet to come."

33 "Tommy" is the slang term for a British soldier. French and German soldiers alike would call for "Tommy" across no-man's land when addressing the British.

to take command.[34]

The Hun who had shot them down wasted no more ammunition on them, after they were once seen to be on fire; but the nose of his plane was always behind their tail. Apparently the gas pressure exhausted itself when they were about 500 meters from the ground, for all at once flickered out the flames. Lieutenant Aignan's furs, uniform and gloves were completely burned off. Though the machine was broken to bits with the bombs on board, they made some sort of landing and saved themselves.

Plotting to Escape

Having tumbled into Bochie without a cent on me, I swapped gloves for a razor with an under jailer. A Belgian boy, whom the Huns found to work about the prison, gave me some previous tips about whom to see if we were able to escape into Belgium. He noted that we lacked a mirror and comb.

The one amusement of the long, dreary day was to peer out through our bars at the half-starved French soldiers who were building railroads and digging trenches between Verdun and us. They were formed into detachments counted and recounted in the little square in front of the prison.

Occasionally we would be taken out for an airing on the battlements to endure the exasperation of watching our own kite balloons for half an hour. The one pleasant incident of these airings was seeing a big Hun biplane make a poor landing on the field to the west of the citadel and smash up. That did our hearts good.

The brilliant offensive of the French at Verdun during the last days of August brought a few more French officers caught in the flux and reflux of the attacks to our prison. The low continuous rumbling that we had been hearing day and night and the bits of news shouted up to our window by newly arrived prisoners had told us that our attack had been carried through with success.

Of course, all of us were keen about getting out, and led by the Frenchman who had come in to my room on the first day, de Malla, we hatched a wild scheme. The jailor guarding our door was to be beaten unconscious with an iron bar we had found. We could get him to open our room door on some pretext. Overpowering him would put us into possession of the front door keys.

34 Second Lieutenant Andre Aignan, Escadrille 111, French Air Service. International Red Cross records show that he was shot down on August 18, 1917. He was flying a Sopwith 1-1/2 Strutter B.2 two-seat bomber with Corporal Paul Boudon as his gunner. Their target was Gesnes-en-Argonne, which was occupied by the Germans at that time. Aignan died on March 28, 1974 at age 85.

Hun Officers Brutal

The descent from the high-walled citadel would have been easy by using all our sheets as ropes and the ruinous moat offered few difficulties. The hard thing would be to cross the line, but de Malla produced a large German staff map which he claimed to have stolen.

We began to smell a rat on account of this last incident, and when on the afternoon before this attempt we were taken out of prison one by one and our clothes and beds searched for the copies of the map we had drawn, our suspicions of this became very definite.

A few days later we were told by a French corporal, who acted as interpreter about the camp, that de Malla had always been there and was certainly in the German pay. We watched very carefully our conversation after that.

The old prison being full, a group of us were taken to a house nearby, which had been converted into a pen. De Malla went with us, but it was evidently seen that his affair was going badly, and one night he disappeared for good.

Our new place proved to be alive with rats and worse, fleas. At night it was a hopeless fight. We finally gave up all attempts to sleep except during the day and talked and sang away the long hours of darkness. Commandant de Bisemont, a splendid type of French active army officer, had a good voice and endless repertoire. He took the lead in keeping things going.

For half an hour every afternoon we walked up and down in front of the house which faced the subterranean casemates under the battlements.

In those damp cellars the Russian prisoner workmen were housed. We would hear them plaintively chanting their melodious folk songs in the evening. They always sang in four-part harmony, and always true to pitch. Perhaps the Russian custom of singing many voiced chants without accompaniment in church had given them their perfect ears and magnificent natural voices.

The doleful wailing of our old Landsturn Guard over their meager allowance of chemical "ersatz" in the canteen across the street indicated as plain as words what those stout missionaries of "Kultur" thought of a beerless campaign.

Another creature whose acquaintance we made was that perfect brute and most efficient slave driver, the German under officer. Our blood boiled to see some of the Russian soldiers, lined up to rigid attention, violently beaten in the face by the fists of the Feldwebels.[35] The provocation offense could not have been great, for it was a daily performance.

"Cold Horse" Sandwich

Some of us found that to carve daggers and napkin rings with our penknives out of hard blocks of oak aided tremendously to pass the time and to keep our thoughts away from the next meal or the banquet we would have when we got back to Paris.

After about three weeks of this life, the longed-for orders to leave for Germany came. We were hauled out of bed at 4 o'clock in the morning and marched down the hill through the silent streets of Montmedy to the station. A reserve lieutenant, who might have been a commercial traveler before the war, with six soldiers, was in charge of 10 of us. The officer found us second-class seats in the train for Thionville, or Diedenhofen, as the Hun called it.

We followed the line of the great railway artery from France to Southern Germany that the American artillery and aviators cut during the last days of the war. It traversed the rich iron and coal region of Briey, where we could see the bleached ruins of the first fights during the Hun triumphant onrush on Paris. Every few minutes we would pass long trains of material and troops. Already the movement back from the Russian front was commencing.

Thionville gave our first experience of being stared at like a strange animal by the railroad station crowd.

No one who has traveled much as a prisoner in Germany will ever again be troubled by stage fright. A few Lorraine-French approached timidly, but did not dare to be seen speaking to us. Finally the crush about us grew to be too much of a good thing and our guardian officer took us to a military waiting room, where we ate a bite of black bread and cold horse.

On the journey to Metz we had several opportunities to talk with French civilians in the smaller stations. They would remark our uniform as we stood in the window of our compartment and come up very boldly to ask, "How things were going? Can one hope for our deliverance in the near future? Are you being well treated?" ("A mais c'est long pour nous autres, messieurs."[36]) Then the station mistress would scold them away.

---

35 Feldwebel is a rank in the German Army equivalent to a U.S. Army Sergeant Major.

36 "A but it's long for us, gentlemen."

The methodical Hun had carefully sorted his work women. The fat ones rustled baggage, the tall, strong ones acted as watchmen and station police while on the trains were the small, agile girls. The latter, clad in knee breeches and long stockings, reminded one of a pony ballet.

The great Metz railroad station was the scene of a most distressing incident for us; our mouths were as dry as dust and we had nothing to drink all morning. The German officer allowed us to order beer and lemonade.

View of the prison camp at Stargard which held British and Serbian prisoners. (Library of Congress)

Maps Secreted

Just as it was about to be served we had to scamper for our train, hissed at on the way by an old German woman. The train was quite full and our ex-drummer had a long argument with the burlesque conductor lady before they attached an extra carriage for us. Drawing out of Metz we got a glimpse of the silhouette of the hills above Pont-à-Mousson, the scene of many rambles in the old American ambulance days in 1915. So near and yet so far with a vengeance.

Everywhere one saw Russian soldiers aiding the peasantry to gather the last hay crop from the fertile prairies or driving the cream-colored oxen home. One easily realizes why the Hun was so greedy to seize this beautiful region, with its varied mineral riches, its fertile bottom lands and its rugged hills clothed with towering conifers.

One of the French officers with us had been captured with the funds of his battalion mess in his pocket. He turned over this money to the commandant, who spent it for ale for us. At way stations the country people were permitted to sell plums to us through the car windows. Anything helped to quiet that craving for something to eat that we had continually suffered since our departure.

While going through the hills between Saarbrücken and Strasbourg, Commandant de Bismont and I made copies of the railroad map in the carriage which I hid behind the glass of our shining mirrors. They did not help us much, for the commandant's was taken in Karlsruhe, and mine in Landshut.

What emotions for those French officers to roll through Strasbourg with quaint Akation houses and its superb Norman Gothic Cathedral! But two of us were back in the French Army when the Tricolor made its entry into the old city.

Rolling across the fertile floors of the Rhine valley in the twilight we reached Karlsruhe in the early evening. One agreeable sight that greeted our eyes were enormous signs, showing people where to go during air raids. It was obvious that since a long time Karlsruhe had had the wind up.

We were interned at the notorious Hotel de l'Europe that practically every Allied officer prisoner has passed through.[37] This place was very cleverly fitted out by the German intelligence for gaining valuable secrets from its involuntary guests. Officers were generally put in rooms alone by themselves for a week and a half and then placed with a friend who had had similar treatment.

The Trick Chamber

After hours and days of silence and loneliness without books, papers or games it is almost impossible to resist the temptation to jabber like a magpie when one is allowed.

All of these rooms were luxuriously equipped with microphones to catch this gossip. Lieutenant Sauvage of the French aviation, who was with us, discovered and tore out two in his chamber. One was hidden in the light-

37 The Hotel Europäischer Hof in Karlsruhe, located a short distance from the train depot, was known by prisoners as the "Listening Hotel" for the amount of eavesdropping devices used there. Located at Ettlinger Strasse 39 in Karlsruhe, today it is the home of the St. Hildegard Youth Hotel. Though the original building has changed considerably in the past century, the entrance to the old Listening Hotel remains the same.

ing fixture – one behind the stream radiator. Of course his confinement was very rigorous after that.

As I could speak English I was put in with Australians, then with Canadians to see if we would compare services and machines. Nearly all of us had already been warned about this place at the front by our captain and took care what he said.

Food in Karlsruhe was a complete change from that at the front. One got an enormous quantity of watery vegetables, carrots, turnips, beets, etc. not too badly cooked. During one-half hour after repast there would be a comfortable tight feeling about the waistline, then one would be hungry again. However, the beds and sheets were clean and we were allowed to bathe once during our stay.

After about two weeks of the Hotel de l'Europe it came my turn to visit the camp commander. As I was the first American through, he kept me for three hours in his office. He said his name was Captain Schmidt and said he was a member of the Reichstag – his good cigarettes and beautiful Alsatian dog is all I can vouch for.[38]

Curiously enough after a few general questions, about why we had come into the war, etc, he started to tell me about Germany and gave me some most interesting sidelights upon the Hun point of view of things, speaking of the violation of Belgium. He said their point of view was as follows:

"A man is attacked in the street (Germany, of course). He has the right to break his way into the nearest house for protection (Belgium). The proprietor of the house has the right to try and prevent his entrance. Any previous agreement between the proprietor of the house and the man attacked in regard to the inviolability of his property is nullified under these circumstances."

Flirting to Europe

Then, of course, the inevitable mutual accusations as to who started the game, the bombardment of towns and killing of prisoners. The end of our interview was punctuated by the alarm for a bombardment by Allied planes.

It was most amusing to watch the antics of the scampering civilians.

38 The Reichstag is the name for the German parliament that existed from German unification in 1871 until 1918. Records indicate that a Robert Schmidt served in the Reichstag from 1893-1898 and again from 1903-1918 as a member of the Social Democratic Party of Germany, and continued in parliament in the Weimar Republic, which succeeded the Reichstag, until 1930. He died in 1943 at age 79. Though not confirmed that this was the camp commander, Robert Schmidt's activity in the Reichstag makes it plausible.

Train motor women left their posts, soldiers broke up their formation to disappear underground like frightened rats. The glass of our windows was painted over to deprive us of the pleasure but we scratched pin-holes which we concealed with soap between times. There were some four raids in my three weeks in Karlsruhe.

Not long before a bomb had landed in the middle of a theatre and had killed hundreds, bringing some of the horror to German hearts that some of us had experienced in the Bar-le-Duc and Nancy bombardments.[39]

The street life was curious. There was a complete absence of automobiles and horses. Frequently we would see dray carts pushed and guided by a dozen children. I have seen officers accompanying their wives to the station – the wives pushing baby carriages containing the officers' valises.

One of the English boys got his guard to unlock his window and carried on a most violent flirtation with a girl in the flat opposite. He fondly had hopes that she might throw him a rope or something. The thing was brought to a sudden close and Captain Schmidt caught his daughter, who was the girl, making signals to some of his own prisoners.

The infantrymen drifted off one-by-one to the camp in Central Park at Karlsruhe, and my orders for a long railroad journey. This time canned horse and black bread was not given to us, instead a paper sack of pieces of sausage, chocolate, sugar and pain d'épices.[40] All soldiers' parcels sent to Southern Germany from France were being seized, and we were profiting by the theft of some poor poliu's parcel.

They were taking me farther and farther from France. Escape became a bigger problem with each mile and my mind simply had to work that much harder.

39 At 3:10 p.m. on June 22, 1916, five twin-engine bi-planes of the French Air Service's Escadrille C.6 hit the city of Karlsruhe with forty bombs. The first five fell next to the tent of the Circus Hagenbeck, which was performing in front of two thousand spectators at the zoological garden across the street from the Listening Hotel. At the end of the attack, 120 people died, including 71 children, and 169 were injured. The Circus Hagenbeck was founded by Lorenz Hagenbeck, the son of Carl Hagenbeck who supplied animals to circuses worldwide, including to Phineas Barnum. Carl Hagenbeck was the founder of the Hagenbeck Circus that toured America in 1903, which was bought out by Benjamin Wallace in 1907, and formed the basis for the Hagenbeck-Wallace Circus. On June 22, 1918, exactly two years after the bombing at Karlsruhe, a Hagenbeck-Wallace train crashed in Hammond, Indiana, which killed 86 people, injured 127 more, and led to the creation of the famed Showman's Rest at Woodlawn Cemetery in Forest Park, Illinois.

40 French spice cake containing rye flour, honey and spices.

MESS AT CAMP VILLINGEN

Photograph sent home, August 13, 1918

Lieutenant Edouard Isaacs (U.S. Navy), Sergeant Harold B. Willis, British officers Jeffrey and Hardesty at the Villingen, Germany, prisoner of war camp, circa late July or early August 1918. Isaacs sent the photograph home on August 13, 1918. Willis and Isaacs escaped from Villingen in October 1918 and together made their way to Switzerland. (Navy Department Library, Washington, D.C./U.S. Naval History and Heritage Command Photograph.)

Chapter Four

The Lights Go Out

Altogether I was in nine prison camps in Germany. Of these I recall the commanders of two as men who showed the human quality. The third camp they sent me to was an enormous barrack near the Dutch frontier, close to the city of Gütersloh. It represented all that was best and most agreeable, and physical pursuits made it possible for one to lead an active life and to forget at times the burden of captivity which constantly overhangs one. Some 2000 officers were kept there. The personality of the commander was the cause of our comparative peace of mind. He was an old Hanovarian army colonel, gruff and crabbed as he could be, yet at bottom fair and willing to grant any reasonable request insofar as the details of our community life were concerned.

We were put first into a quarantine building for several days and the commander came over to look over the newly arrived officers. Someone told him an American officer was there, and he screamed at the top of his voice, "I don't wish to see him; I don't wish to see any Americans." It appears he was much annoyed when America came into the war.

But the comparative comfort of this camp was too good to last. One day the officers received orders to make preparations to leave for a camp in the Province of Schleswig-Holstein, near the city of Lübeck. The camp we were ordered to had never handled so many officer-prisoners together before. Immediately we recognized that we were in for the worst.

The officers in charge of us were typical Prussian brutes. Really, the only word that expresses the Prussian officers is "swine," as the French call them. There was no light in the camp and little water. There were two barracks, into which we were crowded, with a miserable little gravel court for exercise. Absolutely no arrangements whereby we could cook our own food or store it where provided. I have seen poor old French colonels, of three and a half years' captivity, standing in line for an hour to get a miserable plate of cabbage soup. All mental pursuits were quite out of the equation.

Thrust Into Dark Cell

To make things more difficult for us our parcels of food from France and England, on which prisoners altogether depend for food, were stopped throughout the whole month of December and part of January, 1917-1918, and all of us were obliged to eat the musty grain soup, the mussels and fish smelling of pure ammonia, with which the German government supplied us. Many French and Belgian officers protected against the existing conditions, but collective protests were not tolerated by the German authorities, and individual recommendations about almost anything caused their authors to be thrust into a dark cell for a period of from three to nine days. And this was not enough.

Constantly orders came to make our life as miserable as possible, full reprisal conditions. We were crowded into the few rooms so that we had to stack our beds in two or three layers in order to get them in. Exercise and recreation of any kind were strictly forbidden. Our little courtyard was further reduced by a wire fence, so was our allowance of fuel; it had formerly sufficed for three hours a day, now we had a peat fire for only an hour and a half a day. We took turns at this to heat our rice and macaroni in our cups and tins.

Perhaps the most irritating detail of life there were the long, dark evenings, when we sat in our tiny room crowded together on our beds, without light from 4 in the afternoon till we were sleepy enough to retire. We tried giving conferences, lectures, etc., and we discussed the inevitable political questions of the day. Weeks and weeks of this darkness had a most demoralizing effect on us, much more so than the deprivation of washing water, fuel, etc. News of the condition of this camp came to the ears of the authorities in Washington, who, though I wore a French uniform and had no right to the protection of America, came to my aid, through the Spanish Embassy, and I was sent to a very agreeable small camp in Mecklenburg, where I found myself among 120 Russians, two Armenians and one Belgian general.

Prisoner of War camp in Villingen, Germany. (U.S. Air Force)

Treachery Feared

As in every camp, I was the first American there and I was asked by my fellow prisoners to give lecture courses on American institutions and on the English language. This I was glad to do as it contributed to the good cheer. Of course, we were prisoners, and the primal purpose of our life was to get out of prison. We had tried a very elaborate escape, but when we were half-way through the barbed wire fences some of our party were caught, while some succeeded in breaking through. I was among the former, but I was not recognized, so was not under suspicion.

The Russians had already begun to see what the peace of Brest-Litovsk meant and what little hope there was over their freedom under the German heel. Finally, two or three of them signified their readiness to help me in an attempt to make the Dutch frontier with abundant stocks of food. Two Russians a year previous had made a splendid escape from the Mittelberg camp of Bad Stuer. They had got out of the camp by having themselves carried out in the bags of dirty linen, and had actually succeeded in walking some 400 kilometers, carrying all their supplies of food and had reached the Dutch frontier. One crossed the frontier and one was taken on the line. This unfortunate chap was sent to a fortress in Bavaria, from which he again escaped and was later found dead of starvation.

This shows us that, even at the great distance we were from the frontier, there was a good chance of our being successful in getting through. However, as one associates with masses of Russians, the great thing to be feared is treachery, and either through jealousy, or through the Bolshevik element, which is always to be found in German communities, our plans were reported to the commandatur, who searched us thoroughly and found enough on us to confirm their suspicions.

Suddenly, one morning, I was told that, within two hours, I must leave for the other corner of Germany. This was pretty short notice, because, in captivity, one has considerable baggage. At the offices of the commandatur everything was unpacked and most minutely searched. However, as one learns to be crafty when in prison, I succeeded in taking out of the camp my portable wire-cutters, four compasses and my maps. They were hidden inside biscuits, in the thickness of the walls of boxes, inside plaster statuary, etc.[41]

41 While at the camp at Bad Stuer, Willis, who had graduated with a degree in architecture, used his skills to create a plaster bust of the German composer Ludwig van Beethoven. Inside the bust he was able to hide wire cutters and other contraband which he brought to Villingen.

"Swine" the Only Word

This was my third trip across Germany, alone without guards, that is to say accompanied only by a soldier of Hindenburg-Bismarck cast of features who had the stupid habit of many Prussian officers, when angry, of shouting and cursing at the top of his voice, for five or ten minutes at a time, all in German. Sometimes the sense of his scolding is translated by an interpreter, and very often he threatened me with death or the most dreadful punishments if I did not settle down quietly.

The American Red Cross and the American Y.M.C.A. have been doing splendid work for the American prisoners, sending them food, books and athletic games.

Clubbed by Brutes

As we followed up the Neckar Valley, always going South, we made preparations to jump as the train went through a part of the country where few people were about. I did not judge that I had a good chance until I arrived in the city of Villingen, where I was glad to be interned in camp. I am very glad I made no such attempt now, as other American flyers, who had less months of captivity and more energy back of them, did jump in this region, with uniformly unsuccessful and, in some cases, disastrous results. This was a wonderful day to arrive and meet Americans again after 11 months without news of what America was doing. I was astounded at the amount of good American beef the Americans would put away in a day, for during the long months I had passed in prison I had gradually schooled myself to eat only a small morsel of meat a day, at most.

My first impression of Villingen, of the one American officer prison camp there, was of its dirtiness, its primitive sanitary conditions and the pent-in feeling it produced in one. The officers live in the barracks located about a central quadrangle. The spaces between the barracks are filled up with board fences, so that one always has before one the walls of the building opposite, not more than 80 or 90 feet distant. This has a most depressing effect on the prisoners in the camp. If Americans had a long period of captivity to look forward to, as we can be thankful to say they had not, this would have reacted on them in a very depressing way.

The commander of the camp was the typical old Prussian colonel of the "swine" type, hated and feared by everyone with whom he came into contact.

The German, like every bully, is always most disagreeable and high-handed when he is winning. When things are going against him he becomes polite, even makes a pretense of carrying out his agreements and promises, and lies and steals less frequently and when the Allies won their brilliant success all along the line the Hun began to fawn on those he formerly took a delight in ill-treating.

Those of us who escaped hoped that the fact that we broke out of the camp at several points in such large numbers would create such a scandal for the brutish boor of a commandant that he will be returned to the front.

In the camp where I was there were two American officers who had been badly clubbed by the German guards for having jumped from express trains, and who had been re-captured on account of their injuries.

Lieutenant Isaacs of the United States Navy jumped out of the window of an express train. The train was stopped and immediately the guards surrounded him and started in to club him on the head with the butts of their guns.[42] Isaacs, in falling from the train had seriously injured his knees, but the guards forced him to get up and march for six miles and every time they came to a group of civilians they prodded him and beat him and beat him again on the head with their guns, amidst the jeers of the crowd.

A comic feature of the story is that one of the guards was court-martialed for having broken his gun, and as he had broken it on Isaacs' head, Isaacs was called to give evidence at the trial. The same happened to Wardle, who jumped from a train and was clubbed as long as the guards' strength lasted.[43]

The arrival of Lieutenant Isaacs, after the unsuccessful escape, and several American aviators, gave us material to make a serious attempt. For, of course, it was neither duty nor good sense for the American doctors, who were the great majority of the first arrivals in the camp, to be involved in any such action. They were non-combatants and awaited the exchange which was their due from day to day.

---

42 Lieutenant Edouard Isaacs was taken prisoner aboard the U-90 on May 31, 1918, after his troop transport, the *USS President Lincoln*, was sunk by three torpedoes. The *President Lincoln* was built in 1903 by Harland and Wolff, the same company that commenced building the *RMS Titanic* six years later. His guards, Unteroffizier Andreas Lindenberger and Obergefreiter August Moersch, both members of the Landsturm, were outraged at his escape attempt with Moersch clubbing Isaacs with his rifle so hard that the stock broke apart. They then marched him to the POW camp at Villingen.

43 First Lieutenant Herbert Wardle from Memphis, Tennessee, was shot down on June 26, 1918. A member of the 31st Aero Squadron/2nd Air Instructional Center Ferry Command, he was initially held prisoner at Landshut. Like Isaacs, Wardle also attempted to flee the train while in transit to Villingen.

Betrayed by "Friends"

Isaacs immediately spotted the weak point in the defenses. The Germans had constructed a pig sty and rabbit hutch between the back of the stables and the outside wire. A flimsily made barbed-wire fence about 5 feet high was all that prevented a man who had reached the roof of the stable from getting out by crossing the roof of the sty, and jumping off the edge, completely outside of all wire. We had planned to construct a ladder, climb up and lie down on the flat roof of the stable while the inner sentinel was at a distance, jump and scramble over the low wire fence, the same instant that our assistants threw their chains over the electric light wires. Eight of us doing this at once would so startle the outside men of this end that certainly the majority of us stood a good chance for getting away clean.

But Americans had not learned that one should not trust one's best friends in a matter of this kind.

Two days before the night set for the attempt, every American in the camp knew the full details – which worried us not at all. Several, however, made the mistake of telling their best Russian friends, and the rumor spread like wildfire among the Slav part of the camp. Always, as I have noted before, when news of this sort is known by all the Russians it is immediately known by the Germans. Practically all their tunnels in this camp, as in Gütersloh and Bad Stuer, their most ambitious work, were betrayed by anonymous letters the day before their completion.

So it was with us. We were to go Sunday night and Saturday morning a gang of carpenters increased the height of the fence that we were to climb over 20 feet. In addition wire was put around the edge of the roof. The electric lights on that end were made much more powerful. An additional guard was placed inside to watch that special end. The game was up.

Foiled Again

A stupid legal blunder of the Kommandature gave Isaacs and me our next chance. They had adopted a new system of giving word of honor for the walks outside of camp. One wrote on a printed card that one would make no attempt to escape during the promenade, or would aid others in their preparations. This card had our photograph and was stamped several times with the seal of the camp, signed by the camp commander and by ourselves.

At the Bad Stuer two Russians, after the most painstaking work under a microscope, had made a perfect apparent duplicate of their word of-honor cards, but with the wording quite changed, and false signatures and seals. They had escaped quite easily by presenting these cards at the gate, had been retaken and tried for their lives. Death is the only punishment for breaking a word of honor. The German had adjudged that from a legal point of view the German Kommandature was in the wrong.[44] Isaacs and I had comparatively coarse print to copy and we started to work with lenses and inks that I had purchased some time before for the purpose. We told no one, even our bunk mates of what we were doing, but wretched hard luck pursued us again. The cards were finished; duplicate photographs had been bribed from the camp photographer, we were in the act of sand-papering the corners and soiling them, when a Russian turned in the card of his twin brother and was heard of no more. The Huns immediately recognized the weakness of the arrangement and forced us to sign a word-of-honor slip as we passed out of the gate. Again all our preparations were thrown away.

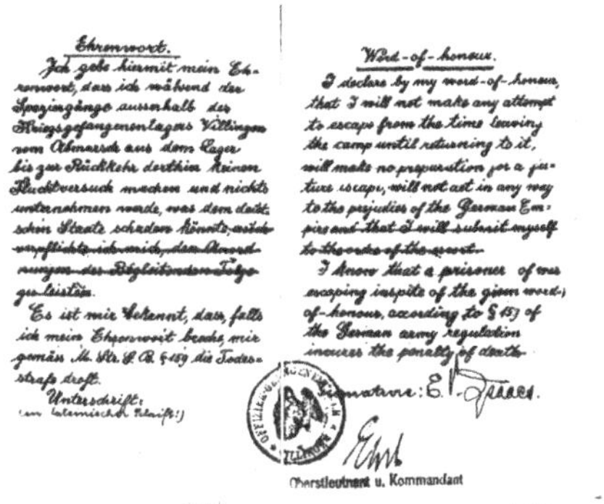
Ehrenwort.
Ich gebe hiermit mein Ehrenwort, dass ich während der Spaziergänge ausserhalb des Kriegsgefangenenlagers Villingen vom Abmarsch aus dem Lager bis zur Rückkehr dorthin keinen Fluchtversuch machen und nichts unternehmen werde, was dem deutschen Staate schaden könnte, ausserdem verpflichte ich mich, den Anordnungen der Begleitenden Folge zu leisten.
Es ist mir bekannt, dass, falls ich mein Ehrenwort breche, mir gemäss M. Str. G. B. § 159 die Todesstrafe droht.
Unterschrift:
(in lateinischer Schrift!)

Word-of-honour.
I declare by my word-of-honour, that I will not make any attempt to escape from the time leaving the camp until returning to it, will make no preparation for a future escape, will not act in any way to the prejudice of the German Empire and that I will submit myself to the orders of the escort.
I know that a prisoner of war escaping inspite of the given word-of-honour, according to § 159 of the German army regulation incurres the penalty of death.
Signature: E. V. Isaacs.

Oberstleutnant u. Kommandant

Lieutenant Isaacs "Word of Honor" card, reproduced from his book Prisoner of the U-90.

Risky, but Funny

We found ourselves anxious that a real try should be made coute qui coute and soon.[45] The best thing looked like trying the opposite small end of the camp from the pig sty. The Huns had a garden between the wire and the buildings here to which access could be gained by a gate fastened by an enormous padlock. With the lights out, a dozen resolute men armed with ladders, rushing out through this gate would have a good chance to climb over the fences and make away. The whole thing hinged on the padlock. The hole was too large and complicated in form to be touched with any keys we could scrape up, getting a good wax impression of the dogs, we made a plaster mould and lead key. It was too soft and wouldn't stand the strain. With another mould, and melting up our belt buckles, buttons and pencil tips, we

44 No American officer ever abused the privilege of the parole walks by trying to escape because of the German concept of high honor among officers. However, the Americans at Villingen used these opportunities to survey potential escape routes away from the camp and often asked their escort to bring them to the south side of the camp, which was in the direction of Switzerland. The parole walks occurred on Mondays, Wednesdays and Fridays and never on rainy days.

45 Coute qui coute is French for "whatever it costs."

cast a brass one. The metal chilled in the mould and could not be gotten to sharply fill the voids. Finally we filed one out of steel but the thing could not be gotten to work.

Fitting these keys was a very amusing performance, it required considerably scenery, for the gate was in plain sight of the inner guard from all parts of the yard from the German soldiers' barracks.

One of the men would bring out a bench and place it approximately in front of the gate, and sit down to read nearby. Others would come out of their blankets, throw them over the bench and chairs and start to pick lice out. Others would follow and take positions where they could watch every German who threatened the security of every man working at the lock. Finally they key-fitter would loaf up, take a seat near the blanket-covered bench, and at a signal from one of his comrades drop behind the blanket and start to work. His hands, which were higher than the back of the bench, were concealed by the head of a man sitting immediately in front of the lock.

"We Must Leave Prison"

You wonder perhaps why we did not try to file a hasp? But it was of such heavy wrought steel that the small files that we had would not have cut it between the time of 5 in the evening, when the gate was closed by the German gardener, and the time when the guards in the interior were doubled and the work would become too dangerous.

While we were fussing with this key one of the Russian officers made a wild attempt at this end, which succeeded from its very foolhardiness. In the full glare of several powerful electric lights he scraped a little hole under the fence, dragging a very crude, long, narrow ladder after him. Within 50 feet of three guards, he sneaked along the ground to the wire, placed his ladder slantingly against the outer wire, scurried up and over. The guard got one shot at him, but missed.

By this time the first dark of the moon in October was half over. Nights were getting cold. Lieutenant Isaacs called a meeting of the hair-brained element that had been working together on these stunts, and stated: "We must leave this camp Sunday night."[46] It was then Saturday morning. That insistence on some sort of action immediately was what got us up.

There was no place for us all to break through together in the same

46 A rumor had swept through camp that 150 Russians would be transferred on Monday October 7, 1918, and the camp would be exclusively American. Isaacs feared that with the Russians gone, the Germans would conduct a thorough search of the camp which would thwart their escape plans.

instant, so we organized three teams.[47] A team captained by Lieutenant Isaacs was to build a ladder out of two strips of wood we used to mark the tennis court with which were long enough to reach from his window sill to the outside wire fence. But in section each piece was only about one by four inches. The thing looked too flimsy for me, and I promised myself that I would not go across that ladder. However, they did very well with it, mounting the thing on Sunday afternoon with screws stolen from every door and window in camp. Material for cross pieces came from Red Cross boxes, bed slats, anything that wasn't nailed down too tightly. A coat of shoe blacking finished it.

Getting the "Makeup"

Major Brown of the American aviation led the second team, and the only infantry officer who happened to be with us.[48] They were equipped with very good wire cutters, which a Russian aviator had given us, and which he purchased the Lord knows where. At zero hour the major's scheme was to go out of the window and cut his way by the fences without regard to sentinels. The last day we invited some aviators, who had just arrived in camp, and who were only four days out of prison for a previous escape, to take advantage of the escape with us. They were only too glad to, as one would expect from an American. They also went out of their windows, and armed with short ladders, climbed over the fences as they came to them. The lieutenant aviator, a Puryear of Nashville, Tenn., the first man over of this team, succeeded in reaching Switzerland alone.[49]

The outfit I led off was composed of three American pursuit aviators. Our idea was to cut our way through the barbed wire fences which separated the prisoner's part of the camp from the German soldiers' part, and go out of the main gate with the guard. I had a knock-down pair of wire cutters that I had made in Bad Stuer to do this work with.

---

47 The various teams had already acted independently in putting together their own escape plans, based upon previous experiences, by the time that Isaacs called them all together to make it a group effort among Americans. Isaacs wanted to escape Saturday night but Puryear's group was already planning for Monday night. They compromised and selected Sunday night instead.

48 Major Harry Milford Brown was commander of the 96th Aero Squadron. On July 10, 1918, six of his squadron's bombers drifted off course and were forced to land behind German lines because of inclement weather. All twelve squadron personnel were taken prisoner and most were in Villingen with Willis. Collectively they are known as the "Lost Flight." Brown was the only American field grade officer at Villingen and had the luxury of his own room with a private bath. The infantry officer was Second Lieutenant George H. Crowns of the U.S. Army's 10th Field Artillery.

49 First Lieutenant George Puryear, 95th Aero Squadron, forced down his first and only German aircraft on July 26, 1918, but they landed behind German lines and he was taken prisoner. A few hours later, after the Germans moved him, the French Army's 63rd Division was in control of the area.

Sunday afternoon was pretty feverish; the first three teams had to cut the wire netting and bars in their windows, while the outside sentinels were at the end of their beats. This was done without detection. Isaacs aided by most of his room was mounting his bridge. Puryear and company were assembling their ladders. We four on the main gate team were stitching together little round red-banded German fatigue hats, making fake buttons for the same, arranging our food about our persons and so forth.

Our friends made blackened wooden guns for us, cut numbers and letters for our coat lapels, made belt buckles and so forth, working like slaves, with considerable risk to themselves, to help us through.

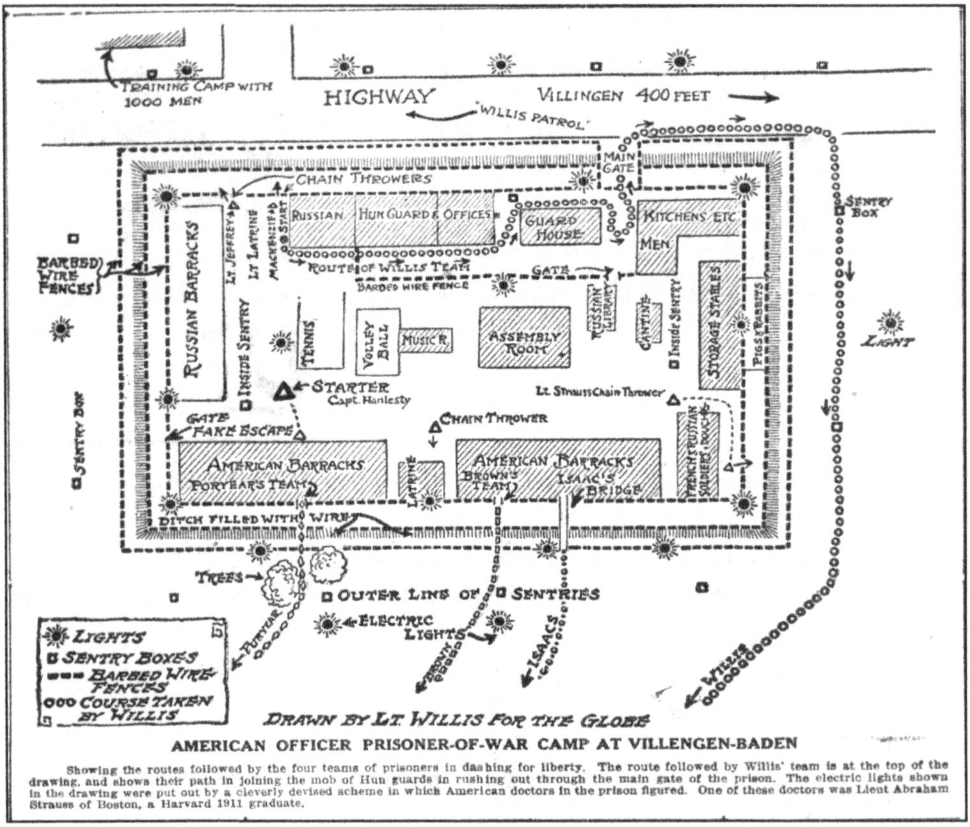

AMERICAN OFFICER PRISONER-OF-WAR CAMP AT VILLENGEN-BADEN

Showing the routes followed by the four teams of prisoners in dashing for liberty. The route followed by Willis' team is at the top of the drawing, and shows their path in joining the mob of Hun guards in rushing out through the main gate of the prison. The electric lights shown in the drawing were put out by a cleverly devised scheme in which American doctors in the prison figured. One of these doctors was Lieut Abraham Strauss of Boston, a Harvard 1911 graduate.

Willis drew this map of their escape route at Villingen, which appeared in the January 12, 1919 edition of the Boston Sunday Globe.

The Lights Go Out

Meanwhile, we had drilled most thoroughly the team that was to throw weighted chains over the various light circuits. The men that were responsible were for attracting the inside guards to out-of-the-way corners. All these people who were not escaping had a chief who gave a signal after a stated moment when the guards were in the best possible position.

Isaacs and I had a light dinner and spent the early evening greasing our bodies all over, rearranging our packs and clothes, and to try and remember that most important article which we knew we would forget.

Ten-thirty, the lights in the room were turned out. The first three teams were in their places ready for action. My lot sauntered out, trying to hide our guns under our coats, and edged over toward the point we were to attack. Ten-thirty-five, zero hour, everybody at their station and quiet. But one of the inner guards in a bad position. The boys watching him must have done their work well, for he soon disappeared.

The boss of the inside men gave a signal! Black night dropped on the camp with the exception of one circuit which flickered on account of the chain swinging. In a few seconds that fuse blew also. When the first circuits went out, in spite of the flickering, Isaacs and Puryear bent back the bars of their window making considerable row.

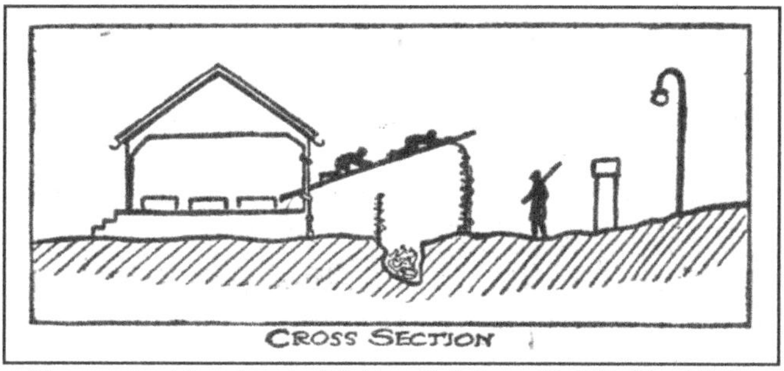

The cross section of the escape route, which appeared in the January 12, 1919 edition of the Boston Sunday Globe. Isaacs group left in this manner.

Thanks to German stupidity, nothing was noticed, while the light on their side was flickering. Isaacs held back the men in charge of sliding on the bridge until complete darkness, then out she went and reached. "Little Tiny Tucker" raced across on his hands and knees, swung off at the feet of two guards facing each other.[50] Isaacs was on his heels and Battle behind him in order of weight. The bridge held like a dream.[51]

---

50 Second Lieutenant Rowan H. Tucker, 96th Aero Squadron.

51 Initial reports in the U.S. newspapers stated that seventy prisoners escaped. There were over seventy American prisoners held at Villingen at that time, but not all of them took part in the attempt. Four groups of Americans are known to have participated in the escape – Willis, along with First Lieutenant Herbert Wardle, 31st Aero Squadron; and First Lieutenant William Chalmers, 94th Aero Squadron; were in one group. Isaacs led Second Lieutenant Rowan Tucker, 96th Aero Squadron; and First Lieutenant Blanchard Battle, 91st Aero Squadron, in a second group. The third group contained First Lieutenant George Puryear, 95th Aero Squadron; and First Lieutenant Caxton Harold Tichenor, 96th Aero Squadron. Major Brown led the fourth group which consisted of fellow 96th Aero Squadron members First Lieutenant Joseph Mellen and Second Lieutenant Alfred Strong. Two additional members, First Lieutenant Carlyle "Dusty" Rhodes, 95th Aero Squadron, and Second Lieutenant George H. Crowns, an artillery officer, were added to Brown's team on Sunday afternoon. Thirteen prisoners total.

Willis as a "German"

The Germans, after the first 40 seconds of stupefaction, started to cry "Halt" and then to fire. During this time Puryear had dropped out of his window and worked his ladder stunt with complete success. He dropped on one of the trees, a Hun on the other side. An absurd game of dodge ensued, which became a bit too hot when a second sentinel approached.

Puryear took to his heels up the sloping meadow. A fusillade on his end of camp commenced. From where I lay on my stomach, quietly cutting my way into the German quarters, it sounded like a tremendous Fourth of July. I had visions of at least half a dozen Americans shot down in their tracks. There is a rumor that Major Brown was killed.[52] That is the only casualty we so far heard of. As the firing commenced to die down a bit the Feldwebbel of the guard rushed in the court of the men where I was lying, shouting, "heraus," "heraus."[53]

The outside guard had already been tripled by the men from the guardroom.

My Huns started to tumble out of bed, grumbling and cursing and reaching for their guns and boots. One, two, three, ran out of the door of their barracks at a jog. I slid along the wall of the building to their doorway and also started off in a dog trot for the guardhouse.

On my path were two of the guard, one so excited that he was shooting into the air.

Arriving at the gate, to find it closed! We had hoped to find everything wide open and the men streaming out on the hot foot. A patrol was being formed and being instructed in the guardroom, where there was a light. I didn't dare go in with my crude disguise. The only thing to do was stall about outside the guardhouse. One of the Huns that I had passed was evidently vaguely suspicious. He came up to me and asked me:

"What are you doing here?"

Fearful of my bad accent, I mumbled something about being half asleep. I had the impression the game was up.

Uphill to Freedom

At that instant the patrol which was being formed in the guardhouse filed out, went to the gate, unlocked it and started out. I sidled over behind

52 Brown survived the escape attempt and was released on November 28, 1918.

53 Heraus is German for "out."

the last man and was out.

My patrol started on a run to the left, a dangerous direction, for it was toward the caserns of the instruction battalion. I cut around to the right and along the back end of the camp behind the pig sties. The guard outside had already been tripled, for where there were posts for two guards I found six. However, holding my gun high over my head, I ran past them without question. On the back corner, with the clear slope of the hill in front of me, I no longer dared keep close to the camp, and cut away at an angle. One of a group of sentinels on the back end challenged me, once, twice, and fired.

That started the fusillade which seemed to keep up about 15 minutes while I dug for it up the slope of that hill, loaded down with heavy underwear and shoes and three or four pounds of food. That was worse than the finish of any crew race.

After a while pounding away up the slope there seemed to be the most curious phosphorescent effect on the ground. Every time I took a leap there was a flash of light in front of my foot. It was puzzling, disconcerting. I could not understand it. Then I thought of my flashlight which was fastened to my belt. I put my hand down, and true enough the jar of the running was causing the light to flicker on and off. More Kamelots Allemonde! What a splendid aiming point for a sentinel behind me. I tore it off and put it in the bottom of my pocket where it could offend no longer.

Finally my wind gave out completely and I went over the crest on a walk, with painfully vivid visions of hundreds of men close behind me.

A peasant's house, Black Forest, Baden, Germany in the 1890s. (Library of Congress)

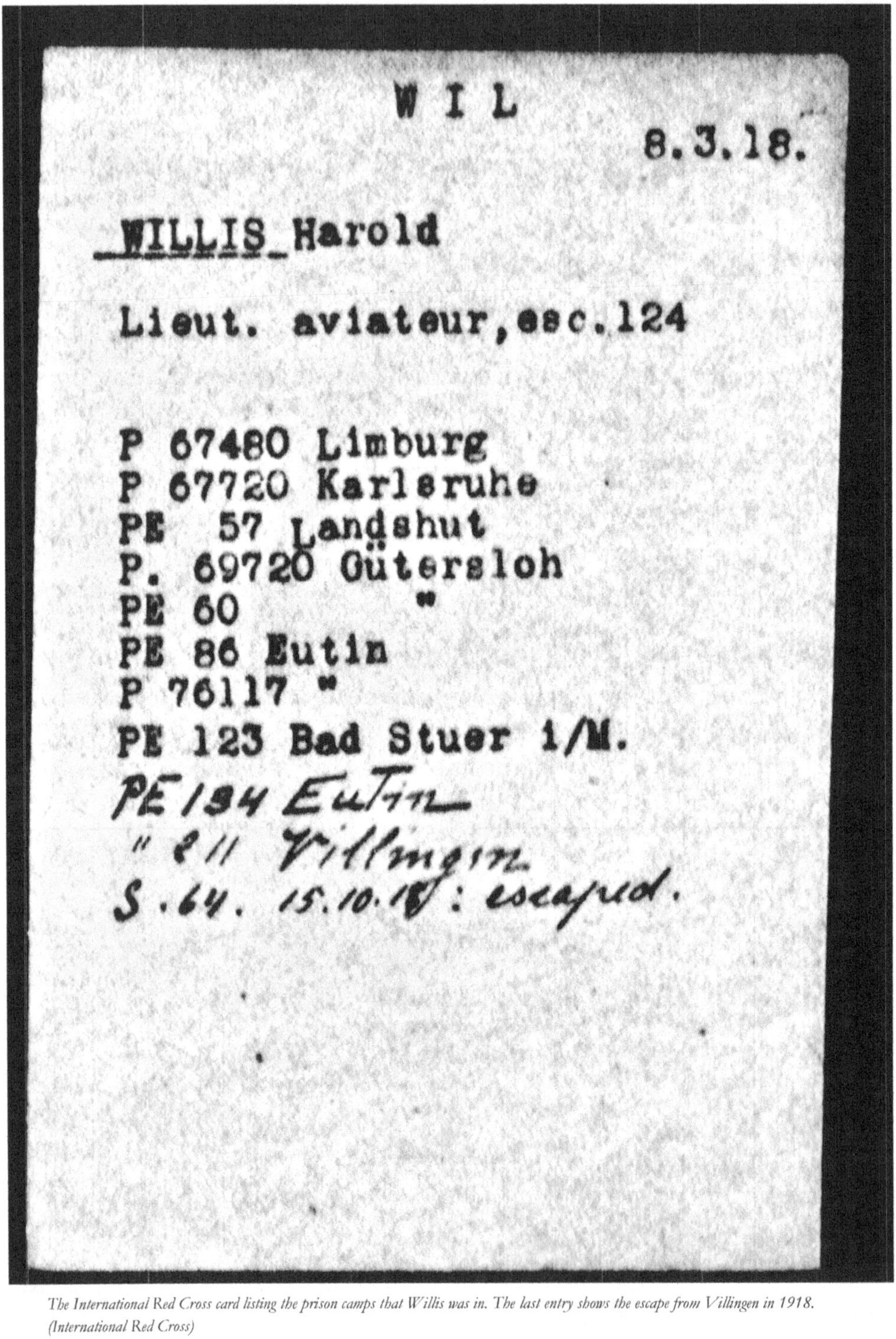

W I L

8.3.18.

WILLIS Harold

Lieut. aviateur, esc. 124

P 67480 Limburg
P 67720 Karlsruhe
PE 57 Landshut
P. 69720 Gütersloh
PE 60 "
PE 86 Eutin
P 76117 "
PE 123 Bad Stuer i/M.
PE 194 Eutin
" 211 Villingen
S. 64. 15.10.18: escaped.

The International Red Cross card listing the prison camps that Willis was in. The last entry shows the escape from Villingen in 1918. (International Red Cross)

Chapter Five

A Night of Risks

Lieutenant Isaacs and I had planned our rendezvous at the foot of a great mound called "Hun's Grave" situated at the top of the second high ridge.[54] Things went well for a while and then we tumbled into our first unforeseen unpleasantness, a swamp. Fortunately there was not much mud. As soon as we could pull ourselves out of one waist-deep water hole we would fall into another. By the time we had blundered out of it, it was a case of being soaked to the skin.

Toiling up the slope, and full of hope of finding him there, I called my name in a low tone. We had agreed that in case of an encounter to speak out own names. There was no reply. I was doomed to continue the long trail alone.

Trying to encourage myself with the thought of the good men that had made much harder and longer journeys without the help of a comrade, I swung on along on the pre-determined route.

Perhaps a mile farther on, as I was skirting the edge of a small wood, I suddenly heard voices close to me. The battalion had cut me off! The ignomy of escaping clear and being recaptured three miles away, I thought. I dived under a young pine tree.

Again the call, but it said, "Isaacs."

We fell on each other's necks. I am sure neither of us had as much joy in meeting anyone as then. He had waited for me at several points along the road and had finally heard the swish of my rubber raincoat behind him. Knowing the Germans did not possess such garments, he took a chance and called.

We went on holding each other's hand for mutual support, when we stepped in ditches.

The first dose of clawing through forests at night was ahead of us. It's no joke. A lumbering road one can follow by watching the split in the skyline overhead and big trees can be avoided by looking up and locating their trunks against the sky. But low brush is hopeless. You push ahead, get your eyes full of twigs, pine needles and finally in desperation you turn around and push yourself through backwards.

Fortunately, we were soon in the open again and skirting the edges

---

54 Isaacs states that "tradition" has it that a Hun leader died and was buried on that spot. When his body was excavated centuries later, the hill formed a huge mound with a hollow crater. Chances are that the "tradition" was made up by a bored prisoner or guard inside Villingen. Nonetheless, a huge mound with a hollow crater would have been a good landmark to rendezvous.

of the woods which we located on our chart. For the first time we saw our pursuers after us. Soldiers on bicycles with bright headlights following along a nearby road. It didn't worry us much for no cycles could follow us across those rough fields.

A Nervous Place

Every time we crossed roads we carefully killed our trail with pepper, but the pursuit was not within hearing.[55] Twice rabbits gave us horrible frights, and a startled deer caused us to think that all was lost for a minute. On the brow of the hills above Tannheim, we took our first breathing spell, nibbled a bit of chocolate.

The mania that the Black Forester has for crowing, ringing clocks helped tremendously to locate villages. The sound of their bell striking the quarter hours enabled us to get a fairly accurate compass bearing of all the unseen villages about us.

By this time we were far enough on our road to risk using a little path through the wood which gave us delightful walking, after clumping through plowed fields. As we approached the more densely populated country about Donaueschingen, the official source of the Danube River, we cut southwest across the Tannheim Brook, and eventually the state road, Brege River and railroad above Wolterdingen.

This latter valley was a nervous place, for it marked an excellent position for them to cut us off. We made it as noiselessly as possible and gave the ground a thorough looking over. Had anyone been there, they would have certainly heard us splash across the power canal and the swift, but shallow river. We didn't see a soul.

The way was clear. A thin layer of clouds over the stars produced a dim luminous light everywhere and the way across country we had so carefully memorized in our walks, easy to follow.

To Bed in a Ditch

An unpleasant 10 minutes with a barking dog, fortunately well tied up, near a forest guard's house, decided us that it was time to make westing deeper into the wooded region, and by picking logging roads through the dense pines, which appeared to follow our compass course most closely, we

---

55 Isaacs explained that pepper was used to mask their scent against pursuing dogs. If a dog encountered pepper, it would impede the scent of the target. This could cause a delay when the guards switched out the dogs before renewing their search.

worked our way to within hearing of the little village of Unterbränd.[56]

We stumbled upon a cross road, but decided that such points should be avoided as dangerous spots where sentinels might be posted. A careful examination of the locality on tiptoe from the edge of the wood showed everything to be safe. Isaacs, standing on my shoulders, was able to read the letters on the sign board, partly by star light, partly by feeling with his fingers. We crawled into the shelter of the woods and under cover of my coat checked up the information we had gained. The needle of light, to which we had cut down our pocket flashlight, worked out splendidly, giving very little glare.

The hours were slipping by; it was time to begin to think about shelter for the day. The village we skirted around to the north, and crossed the Brandbach River into a perfectly impenetrable evergreen swamp. The first gray of dawn forbade us to go farther, though we were pretty close to houses. The trees were so thick and close together that the only way we could make our path into the thicket was by crawling up the bottom of a drainage ditch. It looked pretty safe.

After some discussion we chose a little opening where there was space to lie down and commenced our preparations for the day. First, trousers and underwear came off and were wrung out as best we could; then we very carefully washed each other's feet with the carbolic soap we had brought for that purpose, dried and greased them and put on dry socks. During the whole trip we carefully repeated this performance every morning, which is certainly the reason why we did not develop sore feet, the greatest bugbears of escaping prisoners.

Children and Dogs

Having no coat to put over us we cut a great heap of evergreen branches; crawling under there we cuddled as close together as possible. With our arms about each other we were able to get two or three hours' good sleep before the damp cold of the morning had penetrated to our bones. Then, of course, we both awoke, trembling with cold. Sitting up we tried rocking forward and backward from the hips, which gave us just enough exercise to keep circulation going.

The mist of morning slowly dispersed a few rays of yellow sun gilt at the tops of the trees above us. What wonderful luck! We would really be able to warm and dry ourselves. The warmth crept slowly down through the

56 Westing is a naval navigation term that means "the distance due west made good on any course tending westward."

branches, till finally the long-awaited sunlight touched the bottom of our clearing.

The rest of the day we lolled about on the ground like old cats, only moving when a shadow crept out of us. It would have been a perfect day had not the sound of dogs and children in the woods about us kept our nerves tense most of the time.

Isaacs was sure that these dogs were after us. We had certainly attracted a certain amount of attention at the forester's house five miles back, where the dogs had worried us. It was also certain that the first day after a big break from an officer's camp would see the whole countryside roused and out.

Sometimes the dogs seemed excited about something and approached, then just as we clutched our clubs for action, they would string away to another quarter. Perhaps the last 100 meters up the bed of the ditch saved us, for toward noon the dismal bay died away for good.

Not so the brats, all day long they seemed to be within 100 feet of us, all day long we could distinctly hear their voices, what they said, the sound of their feet as they stepped on dry twigs.

Bucking Through Underbrush

Their labors of stick and acorn gathering did not carry them to where we were hidden, but it worried us tremendously. One of the most peculiar things that we noticed on our little walking trip through the Black Forest was the enormous number of children there seemed to be about. In a marvelous heavenly, wooded country, you might well expect that when you lay hidden in thickets miles from the nearest village or house you would not be bothered by many people passing near you. At least you would not expect to hear children within a couple of hundred feet of you all day long every day you spend crossing the region; we did. I suppose the reason was that the crops were in and the children were sent out to go over every square foot of ground and carry home everything that could be burned or eaten.

With the lengthening shadows and the first chill of the late Autumn evening our suffering from cold re-commenced. One of the principles we decided to hold on to at all costs was to run no unnecessary risk in order to be more comfortable. This included not leaving our shelter of the day until 10:30 in the evening at the earliest, for several of our friends had been recaptured on account of venturing on roads too early in the evening.

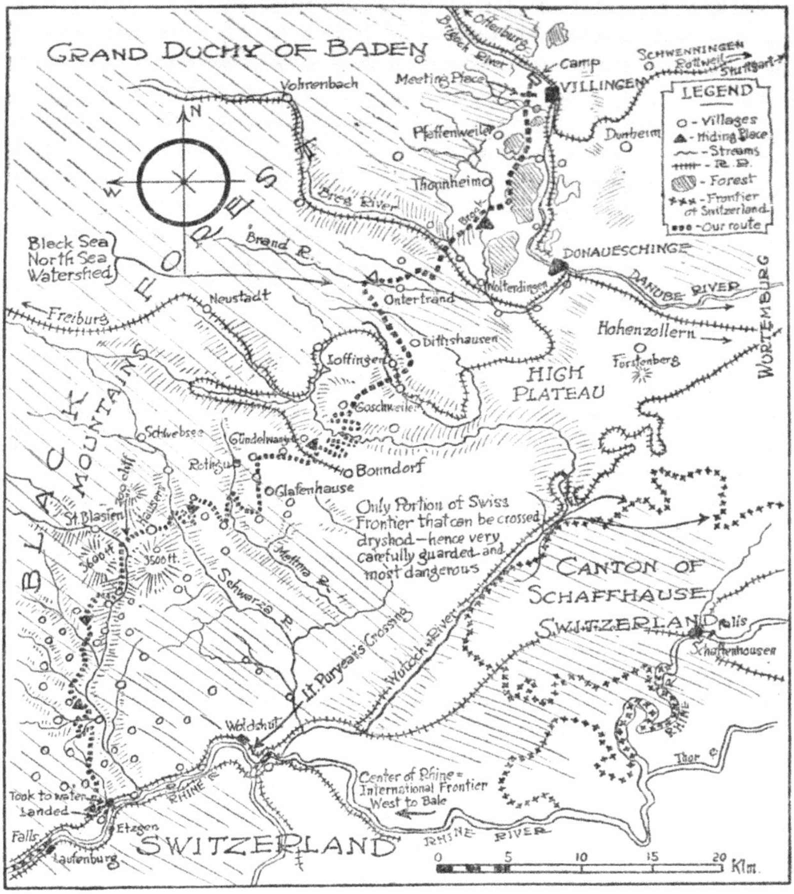

(Drawn by Lieut Willis)

COURSE FOLLOWED BY LIEUT WILLIS AS HE FOUGHT HIS WAY THROUGH THE BLACK FOREST

Map of their escape route through Germany's Black Forest which appeared in the January 19, 1919 edition of the Boston Sunday Globe.

We snuggled up together again as best we could and waited, shaking with cold, for the long early hours of darkness to pass. Finally the moment came when we dared start our night's journey. The first few miles consisted of a compass course directly through the forest. Unfortunately none of the old logging roads led in the right direction; we had to buck our way through a great deal of underbrush and dead trees. By the time we arrived at the wood and roads for which we were aiming Isaacs had lost his hardwood dog protector and I lost something infinitely more important, a good-sized piece of my trousers.[57]

57 Isaacs dog protector was his can of pepper.

Scared of Trees

Things looked comfortably deserted, and we indulged ourselves in the comfort of walking along the edge of the roads for several miles, keeping to the soft ground side in order that we would hear people coming before they would hear us and watching carefully before and behind all the time for bicycles. We did not meet a soul.

We had planned to leave Dittishausen far to our left and strike the railroad somewhere between Rötenbach and Löffingen. After painfully crossing a very deep gravelly ravine we found ourselves above what must have been Dittishausen, for it was a fair-sized town with no railroad. We were so annoyed by the lost mile that we forgot to steal some fine cabbages that we passed.

There was nothing to do but change our route for the entire night and work away around Dittishausen and Löffingen. That was the most exhausting, exasperating, bitter blow that we had encountered on our whole trip. The whole country there was divided up into fields of an acre or two. For generations the peasants had thrown the flints around the borders of these fields, making the dykes of sharp loose stones five or six feet high.

An extra prickly specie of thorn apple had chosen these dykes to grow on most exuberantly. Of course we had to cut across country through these natural hedges. You can imagine the state of our tempers after a couple of miles of this work. Two cedar trees about the size of a man on the edge of the railroad valley gave us a great scare. We waited at least 15 minutes for them to change position. We crossed the open meadow at the valley bottom just north of Seppenhofen, and succeeded in scrambling up the opposite slope without attracting their attention. After a few more thorn hedges we found ourselves in the open and on the road to Göschweiler. Three miles of easy walking brought us to within sound of the clocks of this village.

Plunging Down a Cliff

From all indications on our maps the Wutach River, which barred our route south, was nothing more formidable than brooks we had previously encountered. We had the wrong dope. As we turned south from the edge of Göschweiler the fields started to slope toward the south, at first gently and then pitching off to a 45-degree angle. Anyone who has wandered about much at night knows how difficult it is to dodge heights, depths and angles of slopes by starlight. You cheerfully step off the edge of something and are startled to find yourself up to the knees in mud eight feet lower down, when you confidently expected to find a slight upward slope.

The valley kept caving away before us. Every line of bushes, we would think was a brook, we found to be the edge of a terrace; finally everything pitched down in a wooded slope, which was little less than a cliff, and we could make out dim, mist filled depths below us and the distant roar of violent waters.

A well-graded road to our left seemed to offer the only possibility of going out without risk of a dangerous fall in the dark, although it was strictly against the rules we had set down for ourselves. We looked carefully down, making no noise and listening frequently, but had several soldiers come through there we would have been in a very bad position, with a cliff going up on one side of the road and a sheer drop on the other. Our good luck, held true on the whole venture, is all we can thank for our success. No one was on the road that night.

As we tip-toed our way near to the bottom of the canyon we could see that getting across the stream by wading would be a tough proposition; a great volume of water roared down a deep rock channel with shear sides. Even if it were possible to get down to the water we would be certain to find deep falls that would be fatal, heavily loaded down as we were. We decided to take a chance on the bridge; creeping by the buildings of a water-power plant, we sneaked up toward the edge of the bridge. There did not seem to be anyone on our side. We started to cross, feeling that we were making a mistake and crouched to turn and run if challenged.

A Night of Risks

It was strange that a point like this was not guarded. For a distance of 25 kilometers this almost impassable valley, running at right angles to one's route to Switzerland, has only three ridges. A score of men could have held the whole thing. Further at the east we had been told by a Benedictine monk that all the bridges on the Danube which were located at the same relative position as the Wutach were very carefully guarded. He had told us of one instance he had known about: An English soldier traveling along had been caught on a bridge. He was taken back to the nearest village and put into jail. When they went to feed him he had cut his throat from ear to ear.

As we commenced to breast the dreary switch-back incline up the opposite wall, that towered high above us it commenced to drizzle. Isaacs took the rubber slicker and I buttoned up the collar of my waterproof sheepskin coat, our legs were always wet anyway. Somewhere to the south of Gündelwangen. It was a night of risks. Not 20 minutes after we had crept into cover of the dense foliage we heard woodmen commencing their day's work nearby.

Completely exhausted we ate a piece of chocolate and threw ourselves down on the ground regardless of the drizzle, where in turn we enjoyed an hour of oblivion while the other kept watch. Unfortunately the warmth from the day's ration of chocolate was soon exhausted, and we awoke to teeth-chattering and shaking.

A wood road that followed the railroad toward Bonndorf as it did not appear to be frequently used, we risked following it for several miles that night, crossing the railroad and descending toward the southwest into a deep heavily wooded valley. We looked enough like farmhands not to attract attention, and we could speak German well enough to grunt the "Nacht" that every German always says when he meets another.[58]

In case we ran across a gendarme or soldier who was suspicious and wanted to make trouble the chances would be that he would not be well armed and certainly not as desperate as we were; we always carried two nicely balanced clubs to give one a comfortable feeling of security.[59]

Took to Their Heels

Coming to the last steep slope before the bottom of the valley we saw that there was a large building at the side of the road brilliantly illuminated, throwing great shafts of light across our route; we could not make out what it could be until, coming closer, we discovered that it was an electric plant. We could not hear anyone around, so we simply walked deliberately through the light and passed the place; it was an uncomfortably critical moment. The bridge spanned another roaring torrent – quite impossible to wade as the one of the previous night – some planet or other which had just arisen gave us light enough to see that the place was deserted; we slipped over.

On the other side it was again necessary to pass a building with lighted windows. Thinking that it must be some sort of mill, for it was well toward midnight then, we kept on by. Whatever the place was they were

58 "Nacht" is German for "night."

59 Isaacs and Willis did pass a German civilian that night. They didn't exchange words and the German stepped out of their way. Then the local civilian increased his pace and reported the incident to the police at Faulenfürst. By this time, the two escapees were off the road and traveling through the forest.

supposed to keep watch upon the road. Just as we were by the place a dog started barking, a window slammed open, there was a shout, but the danger was over. We took to our heels up the side of the road and in a few seconds were in the shelter of the woods.

We had a very amusing time going around Rothaus. We thought it was like most Black Forest villages, well-defined groups of houses and fields all about, but this village seemed to have some kind of a suburb. As we were scouting its outlined gardens, bothered by very powerful electric light in the center of the place, we found ourselves all at once walking through the garden passage of a very pretentious house. A lot of hedges and walls made going around the back of the house impracticable, so we kept on through a rose garden into his cabbage patch and down his front walk. Luckily there was not any dog.

Rhine in the Distance

We had no more than reached the Häusern signpost when an enormous black dog dashed out of the village toward us, barking furiously. We were both so startled and frightened that we ran up the northeast road, without bothering about where it led as long as it took us away from that hound. As a matter of fact, it meandered about a good deal over the high plateau, but led us in the right direction.

We circled two or three other small hamlets, dogless, and 5 o'clock found us at a point where the slope of the valley shelved a bit. One could see that there must be a magnificent view there in daytime, for from the heights where we stood as one looked down the great trough of the valley, toward the Rhine, at our left, could be made dimly out against the starry-sky of the highest peaks of the Schwartzwald.[60] But we only stopped long enough to swallow delicious water from a wayside spring and dive into the forest to find a safe thicket for the day.

It was a curious thing that situated where we were, certainly five miles from the nearest village, the woods all about us should be full of children all day long. At one time there was one lone youngster came within a few feet of our tree. Isaacs was down the hill cutting more branches, and fearing that we might make a noise and give himself away, I crept out of the tree on the opposite side and came down to him, for perhaps half an hour that boy kept us crawling about and creeping, and all was ready for a dash;

60 The Schwartzwald is the German name for "Black Forest," and is a wooded mountain range in Baden-Württemberg, southwestern Germany, bordered by the Rhine valley to the west and south. The name Black Forest originates from the general dark color of the numerous pine trees that grow in this region.

finally he wandered off without finding our bed.

On account of such alarms, we always kept our shoes and coats on and our food packed about us, even when sleeping.

A short tramp due west of Häusern the next night soon brought us to a point overlooking St. Blasien and the great horseshoe belt of the Elbe River which was to guide us southward to the Rhine and liberty!

Like Night Animals

Again the brilliant lights of a power plant caused us a great deal of worry. We were coming to have the temperament of night animals. The one place that seemed kindly and friendly toward us was the darkness, the deep woods, the bush. We hated to approach that beam of white light from the open dynamo window as much as a coyote would have done.

Stepping across into an illuminated patch gave us the same sort of shock that an ice-cold shower would have done. Holding our breath, we continued boldly but quietly on. No one had seen us. Fatigue and good luck were making us take risks we had not planned to incur under any circumstances.

The time came for us to quit the valley and take the rough cow trails and steeply sloping banks of the uplands. The climb proved that my legs were not standing up as well as my partner's, for he could take a faster pace with less frequent rests; it was humiliating to be the weaker sister.

Suddenly we heard a great rustle from the brush at the side of the shoot, then a stop, and all was perfectly silent. We froze in our tracks. "Is it an animal or a man?" We stood there holding our breaths for what seemed to be a quarter of an hour – the Thing; Isaacs and I. Again, without any warning, there was a tremendous crash in the undergrowth and the sound of two men running away, on some trail we could not see. They were perhaps more frightened than we were. With a sigh of relief we panted up to the lower edge of the plateau pastures.

So we went on, day after day – huddling in underbrush and trying to sleep by day, and by night pushing through underbrush, slipping around villages, climbing cliffs and plunging through streams.[61]

61 They made their escape on October 6 and it was now five days later. Extreme fatigue had taken over Willis who Isaacs said was "delirious," "babbling incoherently and seemed to have a high fever." He also noted that Willis's hair had "turned almost white" during their week in the Black Forest, Isaacs was still recovering from the beating he received from jumping off the train.

Facing Frontier Guards

The pines and evergreens of this Black Forest were giving away to a greater proportion of beeches, elms and other hard woods. It must be that at last we were approaching the northern slope of the great Rhine Valley. About an hour after midnight one night, while steadily following lumber roads to the southeast, we heard the sound that our ears had been aching for since three days; a distinct whistle and then the unmistakable rumble of a train.[62]

As we had no definite information about how well and where the lines of frontier guards were placed it again seemed the wisest thing to stop. We dove in the woods a ways and started some sort of damp bed. Most of the night was passed rocking forward and back to keep warm. When the blackest of the night began to melt into deep purple and then deep blue we set out again at a rapid clip.

The first people we saw were harmless enough, two shapeless old peasant women in their quaint local costumes. When we discovered them they were ahead of us on our path. We kept far back enough so that we could make out their silhouettes in the mist, but they walked too slowly. A nearby turnip field was not to be resisted, and we made our last vegetable raid; looking out through the foliage we could make out German soldiers pushing milk carts down the road that we were paralleling. They were certainly of the frontier guard.

Eventually our woods tapered down to about one-half acre patch of alder and blackberry bushes. Ahead of us we could dimly make out a great mountain wall, towering up into the mist. Our compasses told us that it ran northeast-southwest, the direction which the Rhine should have at Hauenstein, the point we had been making for.[63]

It must be Switzerland!

Still we had so many times before thought that the frontier was reached that the hill on the horizon must be the land of innkeepers and cheese makers that we did not dare to be certain until we had made sure of the presence of the railroad lines.

Isaacs and I pushed our way into the very southernmost corner bush

62 Willis places this an hour after midnight, but Isaacs recalled it to be "about daybreak on the morning of the sixth day," as the trains did not run past 11 p.m. Because of Willis's sickness, Isaacs account is probably correct.

63 This Hauenstein is a small hamlet two kilometers west of Laufenberg. It is considered to be part of the village of Albbruck in the district of Waldshut, in the state of Baden-Wurttemberg. It is not to be confused with the larger Hauenstein in the Sudwestpfalz district in Rhineland-Palatinate, Germany, and home of the German Shoe Museum. That Hauenstein is nearly 300 kilometers to the northeast.

where we cut out enough brush to give us a place to lie down and construct a sort of a little artillery observation post with the green branches that we had cut, and we took turns watching.

"Promised Land"

Nothing happened about us. Then, way up in the valley, we could make out a moving white streamer. Could it be a train? It was certainly moving positions and coming toward us. There was only one answer: We were there. That train was on a Swiss railroad. Soon after a roar and a white plume of smoke rising out of the deep cut 500 meters to the south of us located for us the German railroad.

Of the river, there was nothing to be seen. The fields ended in a fringe of trees, beyond of which you could make out nothing. Apparently Holstein station was very near to us, for the trains stopped shortly after passing in front of us, and we could distinctly hear the whistle of the conductress, which replaces the American "All aboard" in Bochie.

The one thing that would have been interesting to know would of course been the positions of the sentinels. We saw nothing of these. That night we were to find out why.

Bern, Switzerland's Bärengraben, known locally as "The Bear Pit," as it was in the 1890s. (Library of Congress)

Chapter Six

Long Live the Swiss

A path, a short cut across the fields, passed right in front of our bush about six feet from us. As the day went on a good many people passed by; a short-winded fat woman in a hurry to catch a train, small boys dragging baby carriages full of potatoes, two well-dressed men who dunned an old woman for 40 marks right in front of us. We had filled all the holes of the screen of foliage so that although we could see them, there was small chance that they could see anything of us.

Toward noon the sun came out splendidly. Wet as we were it spelled too much comfort to be resisted; there was a little open place right at the back of our bush where the sun could strike in and we were always protected from the outside by dense bushes. Taking turns watching, we took cat naps there sprawled out in the warm sunlight.

Then occurred what might well have been the disaster of the trip. A farm laborer burst through the bushes so suddenly that we did not have time to move. He threw some implements which he had under a tree, glanced at us without much curiosity and disappeared.

We were thunderstruck, though he had not appeared to be startled. Isaacs was wearing the German hat which I had made because it was more comfortable than his own, and his black coat and trousers vaguely resembled a landsturmer man.[64] Without more ado we sneaked out of the place and back on the road of the early morning, where the woods were fairly wide and we would have a chance for a run for it The fellow had evidently been too stupid to suspect us and there was no pursuit.

Shortly before sunset we ate our last biscuit and chocolate. Then commenced the preparation for the hardest stunt of the whole escape, getting through the frontier.

Modesty or Freedom?

Stripping first one part of our bodies and then another, we greased ourselves thoroughly all over with the lard we had brought for the purpose. I discarded all of my clothing with the exception of trousers, sheepskin coat and socks. Isaacs did the same except that he made a sort of swimming suit

64 The Landsturm was a German reserve force that operated within the borders of Germany. They were not considered to be well trained.

out of his shirt and underwear. We had quite a discussion about that, for it seemed to me that the additional ease that one has with swimming with the bare skin was not worth neglecting. Who cared if we arrived in Switzerland in the state beyond all decencies!

The shoes and extra clothes we hid as best we could. I hope that no Hun has ever found them. The early hours of night until the setting of the new moon were passed cuddled up to each other and rocking as usual. Once the twilight had completely passed, we crawled out into the fields. We had agreed to keep together, not to hesitate to use our clubs where the slightest opportunity occurred and to stick it out whatever happened.

A few steps brought us to a newly plowed field, where we smeared our socks, woolen gloves and faces with mud in order that our color should blend perfectly with the ground. Then we set out for the railroad. For the first 500 yards we were perfectly certain of our going, so we merely took the precaution of keeping of the skylines and taking our precautions. That brought us to the edge of the railroad cut, where the light from the windows of a passing passenger train showed us that there were no fixed guards at least.

We went down through the wire fence and down the embankment to the rock bed railway without too much disturbance.

You cannot imagine how hard it is to cross a railroad line built on this crushed rock bed without making a great noise. The vibration of passing trains prevents the stones from settling into equilibrium and stability. As soon as you put your weight upon them there was a crushing and grinding that could be heard a good way.

Five Seconds Too Late!

You try to put your knee at the end of a tie, and then hold your body up by using the rail as a grip, but in this particular place the stretch was a little bit too far. Getting across as best we could we melted into the bushes on the other slope of the track.

It was not too soon. Five seconds after we had crossed a patrol passed us up the line.

Now was the moment to play slowly and carefully. We had perhaps 300 feet to make on the edge of the railroad track to the jumping-off place we had seen in the morning. We did it on hands and knees, moving slowly and gradually with occasional long waits to listen. We could hear nothing.

A heavy damp fog had formed at the bottom of the valley and the grass was wringing wet. We did not mind much, but by the time we had reached the edge our limbs were soaked through. We peered down through a break in the trees and could make out the edge of the water, which we could hear rustling and slapping against the gravel.

The water was below us 50 feet or so and at a steep angle. What lay between us and there – We listened for what could be below us – What is that? Certainly the sound of a man's foot grinding into the earth.

There again came that same sound from that same quarter. Finally, short, deliberate steps – one-two-three-four-five, and turn, and five steps back.

We located our enemy. But how to get down? Isaacs, being lighter, I let him over the edge down into the darkness.

He found no foothold! It must be a sheer cliff!

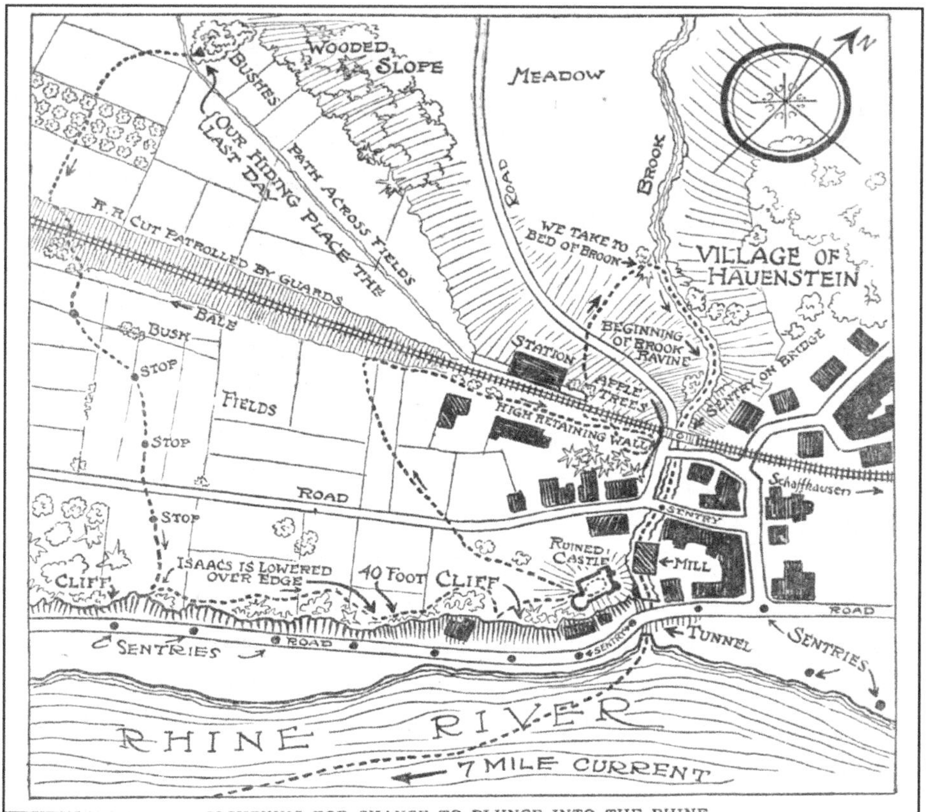

JOCKEYING FOR CHANCE TO PLUNGE INTO THE RHINE

Map drawn for the Globe by Lieut Willis, clearly showing the course taken by him and Lieut Isaacs in trying to break through the cordon of German frontier guards and reach the Rhine. It shows how they were stopped several times by a sheer cliff, again by a searchlight and where they finally slipped into a brook and followed it down to the Rhine.

This map of Hauenstein, Innsbruck, Germany showed where Willis and Isaacs crossed the Rhine. It appeared in the January 26, 1919 edition of the Boston Sunday Globe, the final installment of Willis's six-part series.

Down Into Black Void

We worked along the brink, eastward, letting each other down again and again, and always into that black void. If anything, we were going up – and the cliff was becoming higher.

All this was done so slowly and painstakingly that it was after midnight before we resolved to try something different.

We had found that there was a road at the foot of the cliff where there seemed to be sentinels every hundred meters or so. This sort of thing might continue for miles.

Then we remembered the brook we had followed the previous morning which led under the railroad and through the town of Hauenstein. We struck inland again to the railroad and then worked eastward along its embankment.

Once I touched the signal wires, which twanged and sang for an indefinite time. Isaacs was wild, but he was decent enough not to say what he thought.

We arrived at the railroad station at the outskirts of the village. Every window was dark and everything closed up apparently. A hundred feet beyond the station was the bridge over the gorge of our brook.

That bridge was certainly guarded, so we left the railroad and slid down the slope of the embankment toward the bed of the stream. Suddenly our feet encountered the masonry coping of a retaining wall.

We looked over. There seemed to be a 30-foot drop to the road below. We worked along the edge toward the river, but the situation did not change, and we made our way back toward the bridge again and found ourselves in a hopeless tangle of dead blackberry vines.

Spotted By Searchlight

I happened to be leading. I stepped on a dead branch which snapped under my weight. There was a scurry of feet on the platform of the bridge over our heads and a circle of white light appeared against the slope of the gorge on the opposite side of the bridge.

It studied one place carefully, moved up and down, back and forth slowly.

We did our well-known freezing act for a good quarter of an hour – it was certainly that, for I was standing on one foot, which went to sleep and gave me a terrible time afterward.

Finally darkness closed in on us again. Then we just took French leave of the whole vicinity with as little ceremony as possible and made our way back to the railroad station.

There Isaacs came across an apple tree that he was keen on plundering in spite of the critical state of affairs on this most critical night. He really did not want anything to eat – the idea of depriving Huns of something they were counting on pleased him. Going still farther inland we found a place where the land sloped down to the bed of the stream at a negotiable angle.

Took to the Water

There we took to the water and headed for the Rhine.

Our stream ran over the bedrock, with the occasional loose boulders and patches of gravel. This time there could not be any mistakes. We made every step a study, first feeling with our hands to insure that we would not step on any loose rock which might grind or roll under our weight. The roar and gurgle of the water helped a great deal.

Step by step we made our way back under the railroad bridge, where our friend of the electric light slumbered on undisturbed.

Deeper into a little rock canyon, spanned by the bridges connecting the streets of the town; through the waste tunnel of a town mill and out through the tumbling, rushing waters of its spillway; finally down through a low tunnel, where the bed of the stream tore down a series of great natural steps to the level of the Rhine, into which our torrent slipped in one quiet sweep.

A guard would surely be on the rocky salient that towered above us. Keeping our bodies well under water, we crept down the bed of the stream to where dark, swirling lines and eddies showed its junction with the river. The opposite shore we could dimly make out, for the mist seemed to be less dense on the surface of the water; at least there would be no danger of losing our way in the fog.

We dared not slip off our clothes, for our white skins would have been clearly visible to the eyes above. Looking back I could see Isaacs slipping into the water a few steps behind me. This was to be the last time I saw him that night.

All at once the tug of the river current caught me, rolled me over and over across the shallow ripples at the mouth of the torrent and into deep water. I could slip out of my heavy sheepskin coat quickly, for I had left it fastened with only one button.

Then I struck out. It was like a nightmare. Though I dug away my hardest for the middle of the stream, the current kept swirling me back toward the German shore.

We had been forced to enter the water on the outside of a great bend where all the force of the river was driving in from the center of the river toward the German shore.

I could almost feel whole groups of mitrailleuses trained at the back of my head. The current swept in close under the overhanging bank.[65]

Not a shot was fired.

I made a great deal of a row, blowing and splashing, and I saw that I could not keep on with my heavy, water-soaked silk underclothes on. They were commencing to drag me down.

Free – But No Thrill

While I swam with one hand I snapped my belt buckle and pushed them down over my knees with the other. That was a bad moment.

In those eddies and whirlpools it is impossible to keep above water with one hand, and as I went under those underclothes worried at my feet. I had the distinct impression that it was the last time down.

With that last wild energy of despair I worked them free over my heels and my limbs were free. O! the relief of that cool night air when I at last broke the surface again.

I knew then that I had won!

Swimming easily I touched bottom on the Swiss shore more than a mile below where we had gone in. Where could poor Isaacs be? I shouted, but had no reply. He had not spoken confidently of his swimming – and I felt little hope that he could have made it, hampered as he was.

The bank was steep but not a cliff. I pulled my way up through the bushes to the Swiss railway line, calling from time to time. Where could the Swiss frontier guards be? The place was lonely enough, a railroad line, woods on every side. Nowhere a house! What with worrying about Isaacs and about my own troubles, I confess that I did not feel the thrill of being free again, which perhaps I would have felt in the early months of my captivity.

65 A Mitrailluse is a French volley gun similar to the American Gatling Gun.

Cold, Wet and No Clothes

A cold wind blowing down from the mountains warned me that with a wet skin and no clothes I must keep moving. That morning we had thought we could make out a Swiss town toward the west.

I set out in that direction at a dog trot down the middle of the track. Trap-rock ballast is about the last thing one would pick to run over with bare feet, and after 300 to 400 yards I came to the conclusion that even cross country through the brush was better.

As luck would have it, I found a good smooth highway at about 100 yards inland. Shuffling along for about a kilometer down this road I came to a building; it was a little wayside railway station. There must be someone sleeping upstairs. Banging on the doors and shutters, and shouting, I got no response. With vague ideas of sleeping on the bench inside I tried to get into the place, but it was too well closed.

I started out again and came to a great stone building on the road a short distance farther along. If no one lived there I should have to try and find shelter in the Alps houses. I was bitterly cold in spite of my running and almost "au bout de mes forces."[66]

A beating on the massive front door of the place roused some great dog within which commenced barking ferociously. "Twas the sweetest music I had ever heard; people must be near."

In a few minutes a head in a nightcap appeared at one of the upper windows and demanded "Was ist Los?" – in Swiss German. I told briefly what the trouble was but my chattering teeth were hard to master. The head was sympathetic and hurridly withdrew.

Those good people hurried, I am sure, but it seemed like ages that I waited on that cold stone doorstep.

"Long Live the Swiss!"

At last a young lad unbolted and unchained a great door and let me in. The smell of stale red wine and syrup was unmistakable.

It was a little country inn. I was taken up the back stairs along a long corridor into a clean, roomy bedroom, where they helped me into a feather bed, all wet and muddy as I was.

Then the haus frau appeared and would not let me speak a word until I had spooned down a formidable glass of schnapps. She was a kind,

66 French for "at the end of my strength."

motherly old soul, who took a joy in playing the role of a most efficient Samaritan to me.

On my appeal she packed off the son and heir of the place with a lantern to search the banks for Isaacs. I did not have much hope, though. That done with, the two little strawberry-and-cream daughters appeared with a great tray loaded down with hot milk, chocolate, home-made bread and butter, the former from American flour.

Long live the Swiss!

How rosy and plump and care-free those faces seemed, after the sallow, drawn, careworn countenances of the Huns. The sparkle, the joy of life in their eyes told one that they were proud of being one of the oldest free people of the earth.

May I never again have to live near those fish-eyed Huns.

The folks stayed with me until the boy came back. There was no news of Isaacs!

Before they left the room, I must have been asleep, for the next thing I knew was that someone was shaking me and pulling me violently. Coming to, I saw before me was the plump, jolly visage of some sort of frontier guard. He was grinning from ear-to-ear and apparently babbling over with some good news.

"Ihr Freund its Cezund; he is following me here immediately! We are glad to congratulate you."

It all seemed too good to be true.

True enough, in a moment came Isaacs and the sergeant of the Frontier Guards with an armful of clothes and shoes for me. The sergeant, who was a little jollier and plumper than any of his men, had placed his entire mufti wardrobe at our disposal.[67]

In Queer Costume

Isaacs already had his on. The coat went around him just once and a half, and was pinned under his arms in some way. The wide-brimmed soft hat and flowing necktie, and his rather intense black eyes made him look like a Bolshevik.[68]

Neither of us had suffered any hurt to amount to anything; although

67 "Mufti" refers to the plain clothes worn by a person who wears a uniform for their job.

68 Bolsheviks were a faction of the Russian Social Democratic Party which Vladimir Lenin and Alexander Bogdanov created in 1903. They helped ignite the Russian Revolution in 1917 and changed their name to the Russian Communist Party in 1918. Bolshevik soldiers had a distinct style of tunic and hat for their uniform.

that 200-yard dash on the railroad track the night before kept me from enjoying walking for a while.

My teammate, Isaacs, had about the same experiences as I, but had been carried downstream a little farther. Almost immediately he had come across a Swiss sentinel making his rounds and received the kindliest treatment possible at that moment.

Soon madame, the peaches and cream daughters and a young French-Swiss man joined our circle and prevented my trying on my new clothes. There were a couple of rounds of schnapps and they left me to rig up. Nothing was too small anywhere and what else mattered.

A little while later we bade farewell to those simple, kindly people who would not take a penny of money from us.

The only suit that Isaacs could find to fit him was a very sporty green Tuxedo effect that had something very Boche about its cut. He beat me on the shoes, however, for he found the last good pair of tan shoes in the place. Mine had a sort of an enormous hump, evidently designed to accommodate a bunion.

Land of Treachery

Everyone was mighty decent to us there. We had a couple of dinners with the Swiss officers in charge of the frontier. But they might have given us a room in a safer place. The wall at this point went sheer up from the river, which narrowed down to a rocky gorge leading to the falls a little below. Of course they had to put us in the room nearest Germany – hardly a stone's throw away. We were really nervous.

That gloomy, treacherous land of Germany. Our past week of night traveling made us, and makes me still, think of Germany as the land under a pall of darkness, where the sun does not shine so brightly as elsewhere.

The night in the very comfortable Hotel of Rhinefelden put us in good shape for the journey to the capital, Berne, the next morning.[69]

Here the gendarmes turned us over to the American Embassy and American Red Cross, who squared our expenses up to date.[70]

I am afraid we disturbed those Americans in Berne in the exercise of their duties, but they gave us a royal welcome that went straight to the hearts

69 Rhinefelden is a town halfway between Etzgen, where they crossed the Rhine, and Basel, Switzerland. They spent the night of October 14 in Rhinefelden and arrived in Bern on October 15 where they were turned over to American authorities.

70 Gendarmes are French soldiers or French policemen.

Private Frank Sovicki, Company C, 110th U.S. Infantry, was the first American to successfully escape from a German prison camp into Switzerland. The Shenandoah, Pennsylvania native volunteered for service three days after America declared war. Attached to Company B of the 110th U.S., which was at Hill 204 near Chateau-Thierry on July 14-15, 1918, and cut off from Company C, Sovicki, with two other non-commissioned officers hid in a shell hole. Both his companions were killed by snipers. At night he tried to reach the American lines. Upon seeing a group of soldiers whom he believed were Americans, Sovicki shouted, "Don't shoot, I'm an American." He was captured by the Germans and placed into forced labor with Russian prisoners on a farm near Schaffhausen.

On the night of October 8, 1918, he tricked his guard and locked him up in the guard house. Then Sovicki made a two-day trek for Schauffhausen where the Rhine was little bigger than a brook. He made an unsuccessful attempt to pole vault over the river, then crawled the ten feet to the Swiss side of the river, gaining his freedom on October 11. (American Red Cross photo taken in 1918/Library of Congress)

of three waifs like ourselves.[71]

They took us to dine and to tea to the Kursaal and across the river to see those wonderful bears.[72][73] The American representatives to the German-American conference on prisoners-of-war, which was then framing the international agreement in Berne, did us the honor of receiving us one morning and listening to the stories of our adventures and of the conditions as they actually were in German prison camps.[74]

Headed for the Huns Again

If I remember correctly the latter did not altogether agree with the stories of the German delegates.

There was a funny little Polish soldier from the Pennsylvania mining districts in Berne when we arrived.[75] He had not taken any stock in the Swiss returning prisoners to Germany and so forth. He was going to take a chance anyway. One of his fellow-workmen pointed out a distant mountain top in Switzerland. That was enough, and when no one was looking he dropped his tools and started for that mountain. He must

71 Three waifs refers to Willis, Isaacs and Puryear, who had already arrived in Bern since escaping from Villingen.

72 The Kursaal is a hotel and conference center located in Bern. Construction began in 1859 and it is still operational today.

73 The Aare River flows through Bern. The embassy was located on one bank and the Kursaal and Bärengraben, were located on the opposite bank. Bärengraben, known locally as "The Bear Pit," opened at its present location in 1857 and replaced Bärenplatz, which had opened in 1513. Bärengraben closed in 2009 after a new bear park, BärenPark, opened adjacent to it.

74 In the spring of 1918, the U.S. government proposed holding a conference with the Germans in a neutral country to come to agreement on the treatment of prisoners of war. The conference was held in Bern from September 24 to November 11, 1918.

75 Frank Soviki was the Polish-American prisoner. He joined Willis, Isaacs and Puryear on the rest of the journey to Paris.

have had some experiences, but we could not get much out of him. He was there in Berne, having the time of his life with a permanent grin all over his face.

Four or five days sufficed to have our transports made out and vised, and the four of us started for France.[76] I had a wild idea, more or less supported by railroad schedules, that we could see the Thunaser, the Jungfrau, Mt. Blanc and all Lake Geneva in one day.[77] It was "so American" as the English say. But our main purpose was not sightseeing this time. It was to get back to the front and after the Hun before the season closed.

My schedule had not told about the kinds of trains we were going to ride in, hence at Spiez the boys were not very enthusiastic about following me into a solitary passenger coach at the end of a freight train.[78] The Boston & Maine was a world to be compared with that train! Every time there came a road it stopped and hesitated and walked backward, always to drop cars and take on logs etc.

However, that did not stir our spirits much. We cheered the idle hours by lifting apples from neighboring orchards, much to the brakeman's disgust. The half-grown engine strained on up the Zimmenthal Valley till the grade got too steep for her.[79] Then she laid down and quit…

"Suspicious Characters"

We clambered out, and found ourselves before a diminutive station, amidst magnificent mountain scenery. We were in a grassy basin, with Alpine chalets scattered haphazard about on the sloping green pastures to way up above the snow line. Higher we could make out the outline of a jagged, snow-covered crag that formed a perfect amphitheatre around this. The boys fell on me tooth and nail for leading them up a blind alley.

But no – after nosing about a bit we found two tiny electric cars running on a narrow gauge track, which proved to be the train our timetables called for. Our respect for the line increased when in three hours of strong climbing up the zig-zags to above the snow line the strong-motored little

76 It was actually three days as they left on October 18 for Paris.

77 Thunaser refers to Lake Thun. Jungfrau, at 13,642-feet elevation, is located south of Lake Thun. West of Jungfrau and south of Lake Geneva is Mont Blanc in the French Alps, with an elevation of 15,781 feet.

78 Spiez is a town located on the shore of Lake Thun.

79 The Simmenthal Valley is an alpine valley that goes west from Spiez to Boltigen, and then goes south to Lenk im Simmental. The Simme River runs through the valley.

cars shot us through the last tunnel and out into the warm sunshine high about Montreux.[80]

We shall always remember Switzerland by that wonderful panorama from the tunnel's mouth, where one sweep of the eye carried one down from the basking in the golden sunshine, down to the miniature hotels and cities along the water's edge, across the deep waters to the dark mountain wall opposite, beyond which towered the massive white shoulders of the Dent du Midi at Mont Blanc.[81]

At Lausanne we ran into an influenza epidemic and we were ordered to be quarantined; but we hired a rickety old automobile and dashed with accelerator wide open until we caught a train for the French frontier.

Need one describe the joy of gazing upon the bristling mustache, sparkling eyes and the broad grin of the first French gendarme! That really meant home again, and all one's devoted friends in Paris.

Whiskey plays with Raoul Lufbery while Soda and Douglas MacMonagle watch. (Smithsonian Air and Space Museum Archives)

"Poor Whisky..."

That meant France, really freedom. Those devoted friends in Paris who had made my captivity endurable and my escape possible! Then back to the front beside those few left of the old Escadrille, with the control of a

80 The 1900 *Brief Guide to Bernese Oberland* states, "In Spiez there is a connection between the Lake of Thun Railway and the lines into the Simmen Valley (Spiez-Erlenbach-Zweisimmen), at Zweisimmen transfer to the Montreux-Bernese Oberland line." The Bernese Oberland Railway was known for having a cog-wheel line, which Willis describes in this section.

81 French for "Teeth of Midday," the Dents du Midi is a multi-summited mountain in Switzerland between Lake Geneva and Mont Blanc. It has seven distinct summits.

good Spad between my knees and the shouts of two Vickers sniffing for the wily Hun before me![82]

Although my old friends, comrades and chiefs all gave me a welcome which made me ashamed down inside when I remembered the part sheer fool luck had played in making the difference between the success of my escape and the future of so many others, no reception could be more hearty than that of old Whisky man, the favorite pet and mascot of the old Escadrille.[83]

He would go on growing bigger and stronger and his wife, "Little Soda," did not understand how to play good-naturedly.[84] After having had the run of our mess and our part of the aviation field, the two of them had to be shut up in a miserable old cage in the Jardin des Plantes.[85] Poor old fellow he had never scratched or bitten anyone intentionally in his life. A few days before I was taken, he was gamboling and leaping up on Lufbery's back in the tall grass like a great clumsy kitten, always with jaws that did not squeeze too hard and soft pads.

Poor Whisky, he has had his day.

One afternoon I went down to call on him. He has disagreeable neighbors, a hyena on one side and a panther that snarls at you on the other, but you should have seen him tear around that cage to find a way to come out to me when I gave him Lufbery's call. He quite forgot Soda, who is no longer little Soda, and bowled her over when she got in his way.

Whisky has had many visitors and the guard is indulgent.

When he was a baby he was always quiet if he had a finger to suck; he never lost this habit. Sure enough he took to a forefinger as he had in old puppy days when he was making the acquaintance of a kind of pet not to be found in the Jardin des Plantes. That great rough tongue rolled out and rasped out my entire hand. It was the same old Whisky.

Then he must have his back scratched and his neck tickled and his ears tweated. Finally that great sinewy paw came out through the bars to be shaken and patted. Whisky will never forget his old friends.

82 Vickers is a British company that manufactured the .303 machine guns for the Spad VII aircraft that Willis flew.

83 "Whiskey" was the name of one of two lion mascots purchased by the Lafayette Escadrille. Whiskey was born aboard a vessel that was traveling between Africa and France, and was put up for sale in 1916. Members of the squadron purchased the lion cub. Whiskey developed a fondness for sucking on the fingers of his handlers.

84 "Soda" was the name the second lion mascot.

85 Jardin des Plantes, French for "Garden of the Plants," was established as a medicinal herb garden in 1626, became a botanical garden in 1635, and opened to the public in 1640. The menagerie, or zoo, was added in 1793. Still in operation today, "Menagerie, the Jardin des Plantes Zoo," receives over 500,000 visitors annually and is the second oldest zoo in existence, established 21 years after Vienna's "Tiergarten Schönbrunn."

Photograph of Harold Buckley Willis, sans moustache, taken by an American Red Cross photographer in Bern, Switzerland after his successful escape in October 1918. (Library of Congress)

EPILOGUE

That Such Men Lived

Immediately after the escape, Oberst Otto Ehrt, the camp commandant, initiated a roll call to determine how many prisoners had escaped. In fact, the guards were still shooting when Ehrt gave his order. It was determined that five men – Willis, Isaacs, Puryear, Battle and Tucker – had succeeded in escaping.

Following standard procedure, a search in the area was initiated that covered an area of a two to three mile radius from the camp, while motorcycle couriers and bicycle riders made haste to all the railroad stations in the vicinity. Local police were also notified. These were the bicycle riders that Willis mentioned seeing right after their escape.

Ehrt reported the escape to the XIV Military District headquarters in Freiburg, where further dispatches by telegraph and telephone alerted troops throughout the entire district, with special attention paid to those stationed near the Swiss border.

Wardle and Chalmers

During the escape itself, Wardle and Chalmers, who were part of Willis's group, hesitated when they discovered the main gate was locked. Instead of waiting with Willis for the guards to open the gate, they slipped back through the wire and returned to the compound.

Herbert Allen Wardle was released from Villingen on November 28, 1918 and returned to Memphis. He moved to New York where he married Estelle Ewing Rowe on August 3, 1925. Wardle passed away on August 31, 1973 at age 81 and is buried at the Cedar Lawn Cemetery in East Hampton, New York.

William W. Chalmers was released from Villingen at the same time as Wardle. He married Katherine Williamson on May 14, 1919 and had a career in real estate. Wardle was a member of the National Real Estate Counselors, Rotary Club and Congregational Church of Riverton (N.J.) when he passed away on November 10, 1967 at age 76.

Brown, Rhodes, Crowns, Mellen and Strong

Major Brown and his team of Rhodes, Crowns, Mellen and Strong ran into problems immediately. Their ladder, which was supposed to be used as a bridge to cross the fence, fell into the ditch at an angle. Mellen and Strong had made it outside of the fence and struggled with the ladder, while Brown, Rhodes and Crowns were barking instructions from inside the building. The team gave up their attempt. Mellen and Strong climbed the ladder and crawled back in through the window.

Harry Milford Brown was a rising star in the U.S. Army from the time he graduated from West Point in 1914 until the misfortune that led to his imprisonment. After the capture of the 96th Aero Squadron aircraft and airmen, Colonel William "Billy" Mitchell, wrote in his diary, "The Germans sent back a humorous message, which was dropped on one of our airdromes. It said, 'We thank you for the fine airplanes and equipment which you have sent us; but what shall we do with the major?' I know of no other performance in any air force in the war that was as reprehensible as this. Needless to say we did not reply about the major, as he was better off in Germany at that time than he would have been with us." Brown continued to serve after his November 28, 1918 release from Villingen, but it was evident that his star was tarnished. He resigned from the Army on March 23, 1920, and died in San Francisco, California on March 2, 1960 at age 69.

Carlyle Rhodes, a civil engineer by trade, returned to Lovington, Illinois after his November 28, 1918 release. In 1921, he took a 10-month long trip to Montana and Wyoming, returning in July 1922. Rhodes moved to Payson, Arizona to work in a gold mine, but returned to Lovington to marry Fleta Ascherman in July 1925. He passed away at his home in Phoenix on July 1, 1939 at the age of 45 and is buried at the Keller Cemetery in Lovington, Illinois.

George Crowns was released on November 29, 1918, a day later than most of his comrades. In 1925, he won election for county judge for Kewaunee County, Wisconsin, a post he held until 1944. A 1915 graduate of Marquette University in Milwaukee, where he obtained his law degree, he married Henrietta Kitzerow, who served as a nurse during the war. She passed away in 1930 after a short illness. Crowns went on to become the head of the Catholic Order of Foresters, a fraternal benefit life insurance society, in 1961, a post he held until his death. He passed away in his hotel room at Chicago's Mead Inn and his body was discovered at 8:30 a.m. on

January 23, 1967. He was 78 years old and buried next to his wife at Riverside Cemetery in Nekoosa, Wisconsin.

Joseph Mellen was also released on November 28, 1918. And like Willis, he arrived in France as an American Ambulance driver before the U.S. entry into the war. Mellen married Sylvia Wigglesworth of Garden City, Long Island, New York, at the Cathedral of the Incarnation on June 7, 1919. The marriage was short-lived as she claimed desertion and was granted a divorce in June 1926. He remarried in 1930, this time to Clara Standish Hawkins, a New York buyer for Fortnam and Mason's clothing company. Mellen worked for the Curtiss-Wright Aviation Corporation and later worked for an advertising firm. He passed away on July 4, 1961 at age 67 and is buried in the Forest Grove Cemetery in Augusta, Maine.

Alfred Strong was also released on November 28, 1918. After spending a few days in a French hospital, he returned to Sioux City, Iowa in January 1919. He married Kathleen Trumbauer on February 3, 1923, and served as a Sioux City police judge and an assistant county attorney for Woodbury County, Iowa for many years. Strong left Iowa in 1940 for California and died at the Wadworth Veterans Hospital in Los Angeles on December 31, 1946 at age 51. He is buried at the Los Angeles National Cemetery.

Battle, Tucker, Puryear and Tichener

Puryear and Tichener also used a ladder to get out of their room and, like Brown's team, it fell to the ground. Instead of struggling with it, they leaped out of the window, used the ladder as a running board to cross the ditch, then hoisted it against the gate and made it to the other side.

A guard stood between Puryear and a tree seeming to play hide-and-seek for a few seconds. Isaacs landed nearby. Tucker had already made it out. During the excitement and confusion, Puryear took the opportunity to make a run for it as Battle came over the fence and ran for the trees. The guard, seeing Tichener on the ground, yelled for him to halt. Tichener was escorted back into the camp through the main gate under heavy guard.

Isaacs, Puryear and Battle were relatively close together when they ran up the crown of the hill to the tree line. Isaacs then left the group to rendezvous with Willis. Puryear also changed course to meet up with Tichener, not knowing for sure if he made it out. That left Battle, who tried to meet up with Tucker. They never connected.

Battle set his course for Stühlingen, Germany, where the Wutach River formed the border with Switzerland. He was at the river four days later looking for a place to ford when he was approached by a soldier with a dog. Knowing that he could not outrun either a bullet or a dog, he gave himself up instead of attempting to swim for the Swiss border. He was placed into solitary confinement in Villingen before being transferred to Berlin. The war ended while he was en route, and he was forced to return to Villingen where he was processed out on November 26, 1918. He died on October 13, 1961 in Columbia County, Florida at age 65 and is buried at Arlington National Cemetery. He married Alberta Frances Clarke of Washington, D.C., eighteen years his junior. She died on December 24, 1993 and is buried with him in Arlington.

Tucker, who was supposed to meet Battle, also set a course for Stühlingen. Overcome by fatigue, both mental and physical, he surrendered two miles from the border just northwest of the town. The Germans court-martialed him and were still processing him when the war ended. He was released with the other prisoners at Villingen on November 28. Following the war, he moved to Jackson, Mississippi where he served as chief engineer and assistant manager of the Mississippi State Rating Bureau. He also served as the secretary-treasurer of the Mississippi State Fireman's Association from February 1935 to October 1947. Tucker died on April 17, 1957 at age 63, and his buried next to his wife, Elva Tolles Tucker, in Jackson, Mississippi's Lakewood Memorial Park,

After arriving at the location where he was to meet up with Tichener, Puryear heard about fifty shots fired off. Figuring that Tichener was either captured or killed, he headed due west since he guessed that the search would be in a southerly direction towards the Swiss border, and then would head south towards Waldshut. In fact, Puryear's route paralleled that of Willis and Isaacs, nearly intersecting at Häusern.

At 4 a.m. on the morning of October 11, Puryear crossed the Rhine in a manner similar to what Willis would face a day later five miles downstream. In doing so, Puryear became the first American officer to successfully make his escape from a German prison during World War I. He was taken to Bern the next morning and then met up with Willis and Isaacs three days later for their return to France.

Puryear's story was told to American audiences in a two part series titled, "The Airman's Escape" that appeared in *The Atlantic Monthly* in the April and May 1919 editions. At the time that Puryear's story appeared in print, he was flying in the Far West Flight of the Victory Loan war bond campaign, which brought pilots together to perform airshows in California, Nevada, Oregon, Washington, Utah, Idaho, Montana, Wyoming, Colorado,

Texas and Arizona, in order to induce people to purchase war bonds to pay for the war.

When the campaign concluded, he transferred to the 9th Aero Squadron at San Diego's Rockwell Field. On October 21, 1919 while at Calexico, California, his DeHavilland DH-4 experienced engine trouble upon takeoff. Puryear was approximately two hundred feet off the ground when he banked to the right, lost lift and fell towards the ground. His right wing hit the ground which caused the aircraft to cartwheel. The impact broke a leg and crushed his head. Puryear was dead about five minutes after rescuers pulled him from the wreckage. He was 24 years old. Puryear is buried at the Gallatin City Cemetery in Gallatin, Tennessee.

Isaacs

Isaacs published his memoir, Prisoner of the U-90, in 1919 shortly after his return. It contained an Introduction written by the Secretary of the Navy Josephus Daniels, which may have led to his award of the Medal of Honor, presented by then Assistant Secretary of the Navy Franklin D. Roosevelt on November 11, 1920. Injuries forced Isaacs to retire in 1921. Isaacs was born in Cresco, Iowa, and then raised in South St. Paul, Minnesota, before attending the U.S. Naval Academy preparatory school prior to attending the Academy. Isaacs married Agnes Elmer Cabell, the daughter of Major General DeRosey Carroll Cabell, on June 5, 1915, the day after he graduated from the Naval Academy. Isaacs changed the spelling of his name to Izac in 1925 to honor his ancestral name. He worked for the San Diego Union newspaper until 1928 and, as a Democrat, made an unsuccessful bid for Congress in 1934. He won election to Congress in 1936 where he played a large role in making San Diego a U.S. Navy seaport. After serving for ten years, Izac lost his re-election bid in the 1948 election and left California for a new home in Virginia. He passed away on January 18, 1990 at age 98 and is buried at Arlington National Cemetery.

Willis

After Willis returned to America, he resumed his architectural career and became a partner in the Boston firm of Collens, Willis and Beckonert. His architectural works include Newton (Mass.) City Hall, Riverside Church in New York, along with buildings at Mount Holyoke College, Union Theological School and Springfield College. During World War II, he served in the U.S. Army Air Force in Africa, England and France, retiring as a colonel. His memoir, the subject of this book, was published in a six-part weekly newspaper column in the *Boston Sunday Globe* in December 1918 and January 1919. An abridged version appeared in Volume II of J. Norman Hall's *Lafayette Flying Corps*, published in 1920. Willis married Cornelia Fiske on May 6, 1920. He passed away on April 17, 1962 at age 72, and is buried in Mount Auburn Cemetery in Cambridge, Massachusetts.

That Such Men Lived

Willis, Isaacs and Puryear, though largely forgotten today, reached celebrity status upon their return. Newspapers covered the story of their captures and escape on a regular basis. It was this celebrity status which enabled Puryear to fly in the War Bond tour of the Western United States, put Izac into Congress, and made Willis an in-demand speaker for fraternal organizations throughout Massachusetts.

Then came World War II. With a new generation of warfighters, their exploits of escaping through Germany's Black Forest became yesterday's news. Though Puryear had been gone for awhile by the time World War II began, there is little doubt that their experiences in the "Great War" guided Izac in Congress and Willis in the U.S. Army Air Force during the later war years.

Perhaps it is fitting, then, to recall the words of General George Patton, the U.S. Third Army commander during the Second World War. He said, "It is foolish and wrong to mourn the men who died. Rather, we should thank God that such men lived." Countless numbers of combat veterans, men like Izac, Puryear and Willis, were in mind when he uttered those famous words.

It is my hope that, even after a century, the story of these remarkable men, their tenacity, and their honor, is not forgotten. We need to remember these men as they had lived.

Jeffrey S. Williams
Maplewood, Minnesota
January 15, 2019

Top: Whiskey and Soda with members of the Lafayette Escadrille. Note the aircraft insignia. (US Air Force Photo)

Bottom: Lafayette Escadrille mascots Whiskey and Soda in in their penned enclosure wth an unidentified member. (US Air Force Photo)

About the Authors

Harold B. Willis

A graduate of Harvard University with a degree in architecture, Willis volunteered for service as an ambulance driver in France, flew with the famed Lafayette Escadrille, and resumed his architecture career. During World War II, Willis served in the U.S. Army Air Force, retiring at the rank of colonel. He died in 1962 at age 72.

Jeffrey S. Williams

An Iraq War veteran (US Air Force), Williams graduated with a BA in History and an MBA from Concordia University-St. Paul. He served on the Minnesota Civil War Commemoration Task Force (2011-2015) and is the author of Muskets and Memories: A Modern Man's Journey through the Civil War.

Made in the USA
Monee, IL
08 November 2026

67380497R00059